# Plant Pathology and Plant Diseases

# Plant Pathology and Plant Diseases

Kenzie Cole

MURPHY & MOORE
www.murphy-moorepublishing.com

**Murphy & Moore Publishing,**
1 Rockefeller Plaza,
New York City, NY 10020, USA

Visit us on the World Wide Web at:
www.murphy-moorepublishing.com



ISBN: 978-1-63987-433-0 (Hardback)



**Cataloging-in-Publication Data**

Plant pathology and plant diseases / Kenzie Cole.
p. cm.
Includes bibliographical references and index.
ISBN 978-1-63987-433-0
1. Plant diseases. 2. Plants--Wounds and injuries. 3. Crop losses. 4. Diseased plants.
5. Phytopathogenic microorganisms. 6. Plant pathologists. I. Cole, Kenzie.

SB601 .P53 2022
632--dc23

# Table of Contents

# Preface

The branch of science which deals with the diseases in plants caused by the pathogens and physiological factors is known as plant pathology. This covers various infectious diseases caused by different types of fungi, bacteria, oomycetes, viruses, protozoa, parasitic plants and nematodes. Some of the important aspects of plant pathology are disease cycles, pathogen identification, plant disease epidemiology, economic impacts and plant disease resistance. Pathogenic infections can spread across plants using various mechanisms such as effector proteins, spores, toxins, enzymes, etc. Physiological factors responsible for different types of plant disorders are frost, drought, flooding, nutrient deficiency, deposition of salts, windburns and wildfires. Human activities which result in soil compaction, salination of soil, and pollution of air are also responsible for infections and diseases in plants. This book elucidates the concepts and innovative models around prospective developments with respect to plant pathology and diseases. It is an upcoming field of science that has undergone rapid development over the past few decades. This book will serve as a valuable source of reference for those interested in this field.

A foreword of all chapters of the book is provided below:

Chapter 1- Plant pathology is a branch of biological science that deals with the study of plant diseases, the organisms that cause them and the environmental conditions that facilitate or cause them. A plant pathogen is any organism that causes diseases in plants. This chapter introduces plant pathology in an easy-to-understand manner; Chapter 2- There are a wide variety of pathogens that cause plant diseases. These include the various plant pathogenic bacteria, fungi, viruses, etc. An understanding of these causal agents is critical to the protection and restoration of afflicted plants. This chapter addresses each topic in a detailed manner for a thorough understanding of the subject; Chapter 3- A plant disease refers to any disorder that impairs the normal functioning of a plant or its related systems. Plant diseases are usually infectious and may affect a large plant population if not managed properly. This may result in massive commercial and population losses. This chapter is a compendium of some important plant diseases commonly found in plants; Chapter 4-The reduction of pathogen growth in plants is known as disease resistance. Plant disease management aims to reduce the aesthetic and economic damage caused by plant disease. Plant disease resistance, disease resistance in fruits and vegetables, pest control and fungicide are the important topics covered in this chapter; Chapter 5- The range of adaptations evolved by plants to improve their survival and reproduction are termed as plant defence mechanisms. This chapter thoroughly discusses plant tolerances and defense against herbivory, inducible plant defenses against herbivory and the use of Endophytic fungi by plants in their defense.

At the end, I would like to thank all the people associated with this book devoting their precious time and providing their valuable contributions to this book. I would also like to express my gratitude to my fellow colleagues who encouraged me throughout the process.

**Kenzie Cole**

# Chapter 1

# An Introduction to Plant Pathology

Plant pathology is a branch of biological science that deals with the study of plant diseases, the organisms that cause them and the environmental conditions that facilitate or cause them. A plant pathogen is any organism that causes diseases in plants. This chapter introduces plant pathology in an easy-to-understand manner.

The science, which is concerned with the study of plant diseases and their causes, is known as plant pathology. Plants, like animals and humans, can also become diseased. In fact, there are a much wider number of plant diseases than there are human or animal diseases simply because there are more plant species involved in agriculture, horticulture and forestry than in medical or veterinary medicine.

The science of plant diseases is called plant pathology, or phytopathology. There are a wide variety of microorganisms such as fungi, bacteria, viruses and nematodes that cause these diseases. Diseases caused by these pathogens are often called biotic diseases. Also, environmental conditions, such as winter damage or drought stress, can cause disease in plants. Diseases caused by these factors are often called abiotic diseases.

Plant pathologists are responsible for studying plant diseases, and their studies include various aspects of plant diseases, such as the organisms and environmental conditions that cause disease in plants, the mechanisms by which these factors cause disease, the interactions between these causal agents and the plant, and methods for managing or controlling plant diseases.

The science of plant pathology is closely allied with other sciences such as botany, mycology, microbiology, genetics, chemistry, horticulture, agronomy, and soil science. Plant pathologists integrate and use information from many of these sciences to develop insights into disease development and disease control.

## Plant Disease in the Modern World

Plant pathology has been astoundingly successful in providing farmers with methods for managing plant diseases. This success is one of the reasons why famine in today's world is a relatively rare and isolated event. Nevertheless, plant disease continues to exert a toll on crop and food production.

Diseases cause several billions of dollars in lost production worldwide each year, reducing yields by over 500 million tons annually. These losses reduce profits to farmers and drive up prices of food purchased by consumers. In addition, current disease control measures are in some cases costly and inefficient. Plant pathologists are working around the world to develop new, more efficient, and environmentally sustainable methods to manage plant diseases.

Plant disease is an impairment of the normal state of a plant that interrupts or modifies its vital functions. All species of plants, wild and cultivated alike are subject to disease. Although each species is susceptible to characteristic diseases, these are, in each case, relatively few in numbers. The occurrence and prevalence of plant diseases vary from season to season, depending on the presence of the pathogen, environmental conditions, and the crops and varieties grown. Some plant varieties are particularly subject to outbreaks of diseases while others are more resistant to them.

Blight: Potato leaf infected with a fungal blight.

## Nature and Importance of Plant Diseases

Plant diseases are known from times preceding the earliest writings. Fossil evidence indicates that plants were affected by disease 250 million years ago. The Bible and other early writings mention diseases, such as rusts, mildews, and blights that have caused famine and other drastic changes in the economy of nations since the dawn of recorded history. Other plant disease outbreaks with similar far-reaching effects in more recent times include late blight of potato in Ireland; powdery and downy mildews of grape in France; coffee rust in Ceylon; Fusarium wilts of cotton and flax; southern bacterial wilt of tobacco; Sigatoka leaf spot and Panama disease of banana in Central America; black stem rust of wheat; southern corn leaf blight in the United States; Panama disease of banana in Asia, Australia, and Africa; and coffee rust in Central and South America. Such losses from plant diseases can have a significant economic impact, causing a reduction in income for crop producers and distributors and higher prices for consumers.

Loss of crops from plant diseases may also result in hunger and starvation, especially in less-developed countries where access to disease-control methods is limited and annual losses of 30 to 50 percent are not uncommon for major crops. In some years, losses are much greater, producing catastrophic results for those who depend on the crop for food. Major disease outbreaks among food crops have led to famines and mass migrations throughout history. The devastating outbreak of late blight of potato (caused by the water mold Phytophthora infestans) that began in Europe in 1845 brought about the Great Famine that caused starvation, death, and mass migration of the Irish. Of Ireland's population of more than eight million, approximately one million (about 12.5 percent) died of starvation or famine-related illness, and 1.5 million (almost 19 percent) immigrated, mostly to the United States, as refugees from the destructive blight. This water mold thus had a tremendous influence on economic, political, and cultural development in Europe and the United

States. During World War I, late blight damage to the potato crop in Germany may have helped end the war.

## Diseases: A Normal Part of Nature

Plant diseases are a normal part of nature and one of many ecological factors that help keep the hundreds of thousands of living plants and animals in balance with one another. Plant cells contain special signaling pathways that enhance their defenses against insects, animals, and pathogens. One such example involves a plant hormone called jasmonate (jasmonic acid). In the absence of harmful stimuli, jasmonate binds to special proteins, called JAZ proteins, to regulate plant growth, pollen production, and other processes. In the presence of harmful stimuli, however, jasmonate switches its signaling pathways, shifting instead to directing processes involved in boosting plant defense. Genes that produce jasmonate and JAZ proteins represent potential targets for genetic engineering to produce plant varieties with increased resistance to disease.

Humans have carefully selected and cultivated plants for food, medicine, clothing, shelter, fibre, and beauty for thousands of years. Disease is just one of many hazards that must be considered when plants are taken out of their natural environment and grown in pure stands under what are often abnormal conditions.

Many valuable crop and ornamental plants are very susceptible to disease and would have difficulty surviving in nature without human intervention. Cultivated plants are often more susceptible to disease than are their wild relatives. This is because large numbers of the same species or variety, having a uniform genetic background, are grown close together, sometimes over many thousands of square kilometers. A pathogen may spread rapidly under these conditions.

In general, a plant becomes diseased when it is continuously disturbed by some causal agent that results in an abnormal physiological process that disrupts the plant's normal structure, growth, function, or other activities. This interference with one or more of a plant's essential physiological or biochemical systems elicits characteristic pathological conditions or symptoms.

Plant diseases can be broadly classified according to the nature of their primary causal agent, either infectious or noninfectious. Infectious plant diseases are caused by a pathogenic organism such as a fungus, bacterium, mycoplasma, virus, viroid, nematode, or parasitic flowering plant. An infectious agent is capable of reproducing within or on its host and spreading from one susceptible host to another. Noninfectious plant diseases are caused by unfavorable growing conditions, including extremes of temperature, disadvantageous relationships between moisture and oxygen, toxic substances in the soil or atmosphere, and an excess or deficiency of an essential mineral. Because noninfectious causal agents are not organisms capable of reproducing within a host, they are not transmissible.

In nature, plants may be affected by more than one disease-causing agent at a time. A plant that must contend with a nutrient deficiency or an imbalance between soil moisture and oxygen is often more susceptible to infection by a pathogen, and a plant infected by one pathogen is often prone to invasion by secondary pathogens. The combination of all disease-causing agents that affect a plant make up the disease complex. Knowledge of normal growth habits, varietal characteristics, and normal variability of plants within a species—as these relate to the conditions under which the plants are growing—is required for a disease to be recognized.

## Disease Development and Transmission

### Pathogenesis and Saprogenesis

Pathogenesis is the stage of disease in which the pathogen is in intimate association with living host tissue. Three fairly distinct stages are involved:

- Inoculation: Transfer of the pathogen to the infection court, or area in which invasion of the plant occurs (the infection court may be the unbroken plant surface, a variety of wounds, or natural openings—e.g., stomata [microscopic pores in leaf surfaces], hydathodes [stomata-like openings that secrete water], or lenticels [small openings in tree bark]).
- Incubation: The period of time between the arrival of the pathogen in the infection court and the appearance of symptoms.
- Infection: The appearance of disease symptoms accompanied by the establishment and spread of the pathogen.

One of the important characteristics of pathogenic organisms, in terms of their ability to infect, is virulence. Many different properties of a pathogen contribute to its ability to spread through and to destroy the tissue. Among these virulence factors are toxins that kill cells, enzymes that destroy cell walls, extracellular polysaccharides that block the passage of fluid through the plant system, and substances that interfere with normal cell growth. Not all pathogenic species are equal in virulence—that is, they do not produce the same amounts of the substances that contribute to the invasion and destruction of plant tissue. Also, not all virulence factors are operative in a particular disease. For example, toxins that kill cells are important in necrotic diseases, and enzymes that destroy cell walls play a significant role in soft rot diseases.

Many pathogens, especially among the bacteria and fungi, spend part of their life cycle as pathogens and the remainder as saprotrophs. Saprogenesis is the part of the pathogen's life cycle when it is not in vital association with living host tissue and either continues to grow in dead host tissue or becomes dormant. During this stage, some fungi produce their sexual fruiting bodies; the apple scab (Venturia inaequalis), for example, produces perithecia, flask-shaped spore-producing structures, in fallen apple leaves. Other fungi produce compact resting bodies, such as the sclerotia formed by certain root- and stem-rotting fungi (Rhizoctonia solani and Sclerotinia sclerotiorum) or the ergot fungus (Claviceps purpurea). These resting bodies, which are resistant to extremes in temperature and moisture, enable the pathogen to survive for months or years in soil and plant debris in the absence of a living host.

### Epiphytotics

When the number of individuals a disease affects increases dramatically, it is said to have become epidemic (meaning "on or among people"). A more precise term when speaking of plants, however, is epiphytotic ("on plants"); for animals, the corresponding term is epizootic. In contrast, endemic (enphytotic) diseases occur at relatively constant levels in the same area each year and generally cause little concern.

Epiphytotics affect a high percentage of the host plant population, sometimes across a wide area. They may be mild or destructive and local or regional in occurrence. Epiphytotics result from various combinations of factors, including the right combination of climatic conditions. An epiphytotic may occur when a pathogen is introduced into an area in which it had not previously existed. Examples of this condition include the downy mildews (Sclerospora species) and rusts (Puccinia species) of corn in Africa during the 1950s, the introduction of the coffee rust fungus into Brazil in the 1960s, and the entrance of chestnut blight (Endothia parasitica) into the United States shortly after 1900. Also, when new plant varieties are produced by plant breeders without regard for all enphytotic diseases that occur in the same area to some extent each year (but which are normally of minor importance), some of these varieties may prove very susceptible to previously unimportant pathogens. Examples of this situation include the development of oat varieties with Victoria parentage, which, although highly resistant to rusts (Puccinia graminis avenae and P. coronata avenae) and smuts (Ustilago avenae, U. kolleri), proved very susceptible to Helminthosporium blight (H. victoriae), formerly a minor disease of grasses. The destructiveness of this disease resulted in a major shift of oat varieties in the United States in the mid-1940s. Corn (maize) with male-sterile cytoplasm (i.e., plants with tassels that do not extrude anthers or pollen), grown on 60 million acres (24 million hectares) in the United States, was attacked in 1970 by a virulent new race of the southern corn leaf blight fungus (Helminthosporium maydis race T), resulting in a loss of about 700 million bushels of corn. More recently the new Helminthosporium race was widely disseminated and was reported from most continents. Finally, epiphytotics may occur when host plants are cultivated in large acreages where previously little or no land was devoted to that crop.

Epiphytotics may occur in cycles. When a plant disease first appears in a new area, it may grow rapidly to epiphytotic proportions. In time, the disease wanes, and, unless the host species has been completely wiped out, the disease subsides to a low level of incidence and becomes enphytotic. This balance may change dramatically by conditions that favour a renewed epiphytotic. Among such conditions are weather (primarily temperature and moisture), which may be very favourable for multiplication, spread, and infection by the pathogen; introduction of a new and more susceptible host; development of a very aggressive race of the pathogen; and changes in cultural practices that create a more favourable environment for the pathogen.

## Role of Enzyme and Toxins in Plant Diseases

The term pathogenesis means step by step development of a disease and the chain of events leading to that disease due to a series of changes in the structure and /or function of a cell/tissue/ organ being caused by a microbial, chemical or physical agent. The pathogenesis of a disease is the mechanism by which an etiological factor causes the disease. The term can also be used to describe the development of the disease, such as acute, chronic and recurrent. The word comes from the Greek pathos, "disease", and genesis, "creation". There are several chemical weapons secreted by pathogens that are utilized as they carry out their activities. These weapons include enzymes, toxins, growth regulators and polysaccharides.

### Enzymes

Cutinases, cellulases, pectinases and lignases are often secreted by the pathogenic organism. Fungi, nematodes and bacteria are all known to produce one or more of the above enzymes in specific pathogen-host combinations. Viruses and viroids are generally not considered to secrete enzymes, although some viruses may encapsidate an enzyme in their particle.

Pathogenic organisms either continually secrete enzymes or upon contact with the host plant. The first surface an organism comes into contact with is cuticle and the cell wall of the plant. The cuticle is comprised of a complex wax, cutin, which impregnates the cellulose wall. The cell wall is comprised of cellulose, which makes up the structural framework of the wall, along with the matrix molecules hemicellulose, glycoproteins, pectin and lignin. Thus, penetration into living parenchymatous tissues and degradation of middle lamella is due to the action of one or more enzymes which degrade these chemical substances. Cutinases degrade the cutin on the cuticle layer pre-softening the tissue for mechanical penetration or as a first step in tissue degradation. Studies have shown that several fungi and at least one bacterial species produce cutinases.

Further, evidence indicates that cutinases are continually produced, albeit in low concentrations, with degradation products often inducing even higher levels of cutinase secretion. Studies have shown that in some organims, cutinase production may be linked to virulence. Pectic substances comprise the middle lamella and also form an amorphous gel between the cellulose microfibrils in the primary cell wall. Pectin degrading substances often termed pectinases or pectolytic enzymes include pectin methyl esterases (PME). polygalacturonases (PG) and pectin lyases or transeliminases. Pectin methyl esterases remove small groups such as methyl groups ($CH_3$) often altering solubility and thus affecting the rate of chain splitting by polygalacaturonase and pectin lyase. Polygalacturonases split chains by adding a molecule of water, while pectin lyases split chains by removing a water molecule from the linkage. Pectin degrading enzymes are involved in a wide range of plant diseases particularly in the soft rot diseases (Examples: bacterial soft rot due to Erwinia carotovora, leaf drop of lettuce and water soft rot of crucifers (Sclerotinia sclerotiorum) and damping off in various seedlings due to Rhizoctonia solani). Organisms such as these, as well as, elaborate pectic enzymes lead to tissue maceration. In fact, these enzymes are sometimes referred to as macerating enzymes.

## Cellulases

Cellulose is the major framework molecule of the plant cell wall existing as microfibrils with matrix molecules (glycoproteins, hemicelluloses, pectins, lignins) filling the spaces between the microfibrils and cellulose chains. Cellulases have been shown to be produced by many pathogenic fungi, bacteria and nematodes. Cellulolytic enzymes play a role in softening and disintegration of cell walls. No doubt cellulolytic enzymes are involved in the invasion and spread of the pathogen, but also are instrumental in the collapse of cells and tissues. As pointed out in Agrios text, indirectly cellulolytic enzymes participate indirectly in disease development by releasing soluble sugars that may be used as nutrients by pathogens and also may be involved in the release of materials in the vascular system interfering with transport or translocation of water.

## Hemicellulases

Hemicelluloses are complex polysaccharide polymers that link the ends of pectic compounds to cellulose microfibrils. Since hemicelluloses are such a diverse group of polymers such as xyloglucans, glucomannans, glactomannans, arabinoglucans, etc., several hemicellulases have been identified in many plant pathogenic fungi. The mechanism by which they participate in cell wall breakdown is not clear, nor is it known how they contribute to pathogenesis.

## Ligninases

Lignin is a phenylpropanoid which is found in the middle lamella and secondary cell wall of plants. More than anything else, lignin confers the tough, woody nature to woody tissues. Only about 500 species of fungi are capable of decomposing wood. Most lignin degradation is by basidiomycetes known as white- rot fungi. These fungi produce ligninases that enable the fungi to utilize lignin.

## Toxins as Chemical Weapons of Pathogens

Toxins have been implicated in plant disease as far back as deBary who advanced a theory of plant disease often termed the "toxin theory". A primary tenant of the toxin theory is that a toxin elaborated by a pathogen may produce all of the symptoms of the disease. As more information was developed the theory was largely discarded. The discovery of the toxin victorin, a host specific toxin, revived interest in the toxin theory of plant disease. Toxins may act directly on living host cells, damaging or even killing the host. Some toxins are active on a wide range of plant species (non-host-specific) or in some cases, as with the toxin victorin (host-specific). Non-Host Specific Toxins:

- Tabtoxin: Pseudomonas syringae p.v. tabaci.
- Phaseolotoxin: Pseudomonas syringae p.v. phaseolicola.
- Tentoxin: Alternaria alternate.

## Tabtoxin

This is the toxin involved in the "wildfire disease of tobacco". In this disease, leaves exhibit necrotic spots surrounded by a yellow halo. Identical symptoms of the disease may be induced by culture filtrates of the organism or purified toxins with symptoms identical to that of wildfire of tobacco. Similar effects may also be observed on a relatively wide range of hosts, thus making the toxin non-host specific. Tabtoxin chemically is a dipeptide composed of the amino acids threonine and tabtoxinine. In the cell tabtoxin is cleaved releasing the tabtoxinine moiety which is the active toxin. The inhibition of the enzyme glutamine synthetase is the primary mode of action of the toxin.

## Phaseolotoxin

This is the toxin involved in one of the bacterial bean blights called "halo blight". Symptoms of the disease incited by the bacterium can be produced by the toxin alone. Chemically the toxin is a tripeptide of ornithine-alanine-arginine with a phosphosulfinyl group. Within cells the toxin is enzymatically cleaved releasing phosphosulfinylornithine which is the toxic moiety. Cellular affects are a result of the inactivation of the enzyme ornithine carbamoyltransferase.

## Tentoxin

This is the toxin produced by Alternaria alternata. The disease induced by this organism is primarily a seedling disease in a wide range of plant species. Seedling death results when greater than one-third of the leaf area become chlorotic, and reduce vigor with less than that amount of leaf chlorosis. The toxin is a cyclic tripeptide that binds to and inactivates a chloroplast-coupling factor protein involved in energy transfer and also the inhibition of light dependent phosphorylation of ADP to form ATP.

## Host-Specific Toxins

The idea that toxins play causal roles in plant diseases is attractive and dates back to the time of deBary. When one reduces the toxin theory to its most elementary form, it can be stated that all of the symptoms of a given disease result from the direct action of a toxic product of the pathogen in that disease. Tenents of this theory can be tested by the following criteria:

- The toxin, applied at concentrations which could be reasonably expected in or around the diseased plant, produces in a susceptible host all the symptoms characteristic of the disease.
- The pathogen and the toxin exhibit similar suspect (host) specificity.
- The ability of the pathogen to produce the toxin varies directly with its ability to cause disease.
- A single toxin is involved.

The first toxin isolated that fulfilled the above criteria was that produced by Cochliobolus (Helminthosporium) victoriae. "Helminthosporium leaf blight of oats appeared in 1945 with the introduction and widespread use of the oat variety Victoria' and its derivatives which contained the Vb gene for resistance to the crown rust disease. In the victoria lines, the infects the basal portions of plants, and as it infects, the fungus elaborates a toxin that is carried to the leaves, causing a severe leaf blight, and death of the plant. All other oat varieties and other plant species are either immune or sensitivity to the toxin was proportional to their susceptibility to the fungus. In addition to the toxin producing external symptoms identical to that induced by the pathogen, internal biochemical and histochemical induced by the toxin are identical to that induced by the pathogen.

Chemically, the toxin is a complex chlorinated cyclic pentapeptide. Ultrastructure studies demonstrate the primary target of the toxin is the plasma membrane. Here, it apparently binds to proteins and in some yet to be determined mechanism alters the metabolism of susceptible cells.

## T-Toxin

This is another good example of a host-specific toxin. The toxin is produced in common corn disease "Southern Corn Leaf Blight" incited by the fungus Cochliobolus heterostrophus formerly known as Helminthosporium maidis. T-toxin is produced by race T of the fungus that first appeared in the United States in 1968. By 1970, severe losses occured throughout the cornbelt in corn that carried the Texas male-sterile (Tms) cytoplasm. Corn varieties with normal cytoplasm were found to be resistant to the fungus and also to the toxin. The toxin, chemically, is a mixture of linear, long polyketols with 35 to 45 carbon atoms. The toxin specifically affects mitochondria of susceptible cells where ATP synthesis is inhibited.

## Hc-Toxin

One other example of a host-specific toxin in the HC-Toxin produced by the fungus Cochliobolus (Helminthosporium) carbonum which incites a leaf spot disease in corn. The toxin is specific to only certain maize (corn) lines. The mechanism of action of the toxin is poorly understood at the present time, although the molecular and biochemical basis for resistance is well understood.

## Growth Regulators as Weapons

As you know, plant growth is the result of three factors: cell divisions, cell elongation and cell differentiation. Plant growth is controlled by a group of naturally occurring compounds with hormonal action that are often referred to as plant growth regulators, plant growth hormones, etc. The major groups of plant growth regulators are the auxins, gibberellins, kinins and ethylene. Plant growth regulators are not unlike any hormone, in that, they act is very small concentrations, with slight changes in the normal level resulting in profound changes in the growth pattern of the plant. Pathogens often cause an imbalance in the hormonal system by causing the infected plant to produce more or less hormone, or in some cases the pathogen itself elaborates hormone thus changing the hormone level. Some of the commonly observed symptoms related to effects on plant growth regulation are: stunting, overgrowths, galling, root branching, adventitious root formation, defoliation, resetting, leaf epinasty, etc.

## Auxins

Auxins are the major growth regulator in plants and are defined as growth regulators whose major mode of action is cell elongation; they resemble indole-3-acetic acid in their activity. The naturally occurring auxin is indole acetic acid (IAA). Increased levels of auxin occur in plants infected with fungi, bacteria, viruses, molicutes and nematodes. Following are a few important diseases directly related to altered auxin levels: Corn smut (Ustilago maydis), clubroot of crucifers (Plasmodiophora brassicae), and southern bacterial wilt (Rolstonia solanacearum formerly Pseudomonas solanacearum, crown gall (Agrobacterium tumefaciens) good examples of diseases where altered auxin levels have been implicated in disease.

## Gibberellins

A century ago, rice farmer in Asia noticed some exceptionally tall seedlings growing in their paddies. Before these rice seedlings could mature and flower, they grew so tall and spindly that they toppled over. In Japan, this aberration in growth pattern became known as bakanae (`foolish seedling disease') disease of rice. In 1926, Kurosawa, a Japanese scientist discovered that the disease was caused by a fungal pathogen, Gibberella fujikuroi. By the 1930s, Japanese scientists had determined that fungus produced hyper-elongation of rice stems by secreting a chemical, which was given the name gibberellin. Gibberellins are normal constituents of green plants and also produced by several other microorganisms. The best known gibberellin is gibberellic acid. In the past years, scientists have identified more than 80 different gibberellins, many of them occurring naturally in plants. Spraying of diseased plants with gibberellin overcomes some of the symptoms (stunting) caused by several virus- or mollicute-pathogens indicating that gibberellin involves in disease development.

## Ethylene

Ethylene production in infected tissues can be dramatically induced. This induction is largely dependent on activation of the ethylene biosynthetic pathway in plant tissues. Genes encoding several key enzymes involved in the ethylene biosynthesis are highly activated at the transcriptional level. It has not been shown that ethylene is produced directly by plant pathogenic fungi and bacteria.

Ethylene has been considered as a signal in plant for wounding and senesces responses. Ethylene together with another signal component jasmonic acid may play an essential role in plant defense responses of several pathosystems.

- Induced ethylene production in infected tissues.
- Ethylene induces expression of certain types of PR-genes and some signal components involved in defense signaling.
- Ethylene insensitive mutants block several PR-gene-expressions.
- Ethylene insensitive mutants can be beneficial against some pathogens but deleterious to resistance against other pathogens in a specific a gene-for-gene manner.

Other chemical weapons include polysaccharides, plant defense suppressors, transporters etc. Plant viruses and viroids are not known to produce any substances themselves, but they adapt, induce and manipulate the host metabolism to replicate themselves.

## Importance of Plant Pathology

Plant Pathology has advanced techniques to protect crops from losses due to diseases. The science of plant pathology has contributed disease free certified seed production. Most of the diseases with known disease cycle can now be avoided by the modification of cultural practices. With the knowledge of mode of disease spread, many diseases of economic importance can now be checked, minimized or controlled.

Crop improvement and varietal resistance have been achieved against many diseases through the joint effort of breeder and plant pathologist. Plant pathology has made possible to restrict the spread of plant diseases from one place to other and one country to other through suitable measures and quarantine legislation. With the knowledge of studying about plant diseases, various prophylactic measures are adopted for successful management of diseases. These measures are seed treatment in seed borne diseases, soil treatment and crop rotations.

Diseases can be avoided in cold storage by the application of plant pathological measures as per recommendations made for different diseases for protection of fruits and vegetables in storage. Plant pathology has made possible to recognize, discard or utilize toxic substances by producing toxic substances or by competition or by parasitism. Organism exerting such lethal or damaging effect on the other is called antagonist.

## Principle of Plant Pathology

There are a large number of guiding principles in Plant Pathology, which are often difficult to formulate because biology has so few absolutes. Here are 10 principles, which may aid in understanding, diagnosing and controlling diseases of ornamental plants:

Principle I: Disease is a malfunctioning of a plant, which results from a continuous irritant by a pathogenic agent. This definition of plant disease includes organismal causal agents which attack plants such as fungi, bacteria, mycoplasma, viruses, nematodes and parasitic plants. However, for ornamental plants abiotic diseases far outweigh the traditional causes of plant diseases. For ornamental plants the Plant Pathologist must first examine such abiotic diseases as: 1) high or low temperatures, 2) excess or deficiency of water, 3) excess or deficiency of light, 4) lack of oxygen, 5) air pollution, 6) nutrient deficiencies, 7) mineral toxicities, 8) soil alkalinity or acidity, 9) toxicity of pesticides, 10) improper cultural practices and 11) lightening, before suspecting that the disease is caused by an organism. In many cases "diseases" which apply to ornamental or backyard plants may not be diseases at all since they do not cause a malfunction of the plant but instead cause an unpleasant response on behalf of the home owner. Examples of these "diseases" may be slime mold fungi, which cover plants or fairy rings, which cause the lawn to be unsightly. So "plant disease" often depends on your perspective.

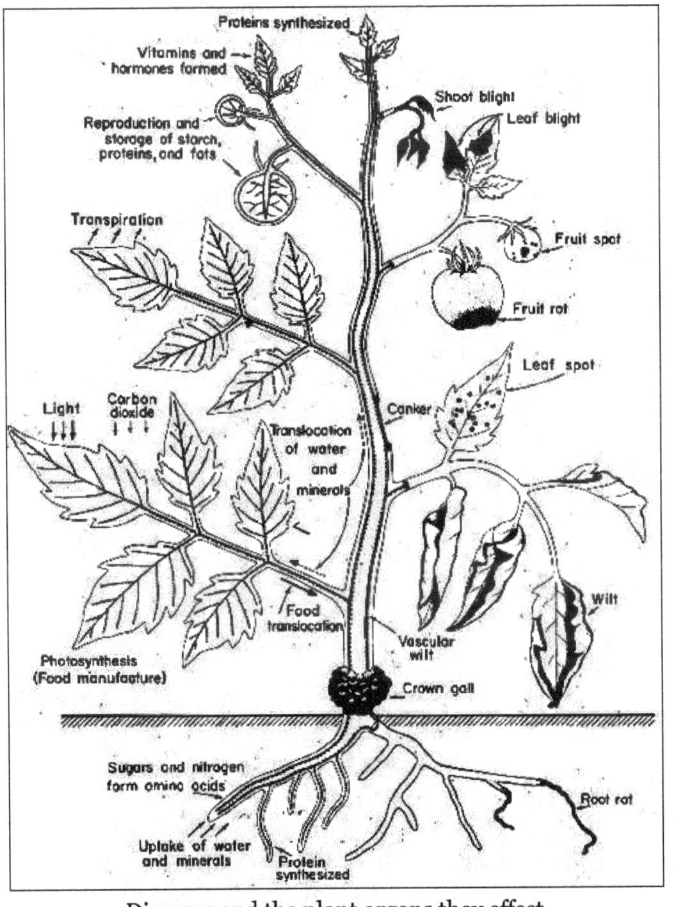


Diseases and the plant organs they effect.

How the disease affects a plant and what organs it causes to malfunction can aid in the diagnoses and control of plant diseases. Leaf diseases are commonly associated with necrotic spots and/or signs of the disease such as spores or fruiting bodies of the attacking fungus. A common mistake made by homeowners is to assume that dying leaves are caused by leaf diseases. More often than not if the plant suddenly wilts or shows symptoms on all leaves the disease is caused by either a blockage in the water conducting vessels, which is called a "wilt disease" or it is caused by a "root rot disease". These two types of disease are often difficult to separate simply by viewing the gross symptoms of the plant. The plant stem must be severed to look for telltale stains in the stem caused by wilt diseases and the roots must be examined for symptoms of root decay. Cankers are a fourth type of disease which are usually characterized by conspicuous lesions on the stem or branches. This type of disease does not interfere with the water movement in a plant but with the movement of sugars and other substances in the phloem. With this rudimentary knowledge of the disease it is possible to narrow your diagnosis to the plant organs which are affected. A search of these organs for further symptoms or signs of disease will usually result in a proper disease diagnosis. However, isolation or identification of biotic pathogens is required for a final diagnosis.

Principle II: Disease results from an interaction of the virulence of the pathogen, susceptibility of the host, and the conduciveness of the environment. The concept that plant disease is not caused by a single organism or disease agent is overlooked by most people. Plant disease is instead caused by a combination of three factors: 1) a pathogen, 2) environmental conditions, and 3) host response. This concept is visualized by the "disease triangle" in which the three factors, pathogen, environment and host make up the three sides of a triangle and the area of the triangle is the amount of disease produced in the plant. The disease triangle shows visually that a very weak and inefficient pathogen could cause substantial disease if the environment is conducive or if the host is extremely susceptible.'

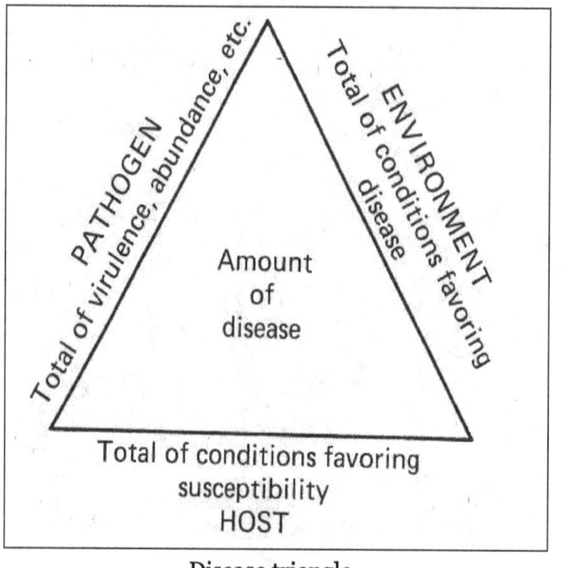


Disease triangle.

Principle III: Conditions which favor plant growth and health commonly favor disease. One of the most common misconceptions among horticulturists and the public is that healthy plants are somehow more resistant to plant disease and so abundant water and fertilizer are often treated as pesticides to insure plant health. While this may be true for weak or opportunistic plant pathogens it is not true for most competent pathogens. For example, high levels of nitrogen fertilizers, which result in rapid growth of plants, is often prescribed to improve plant health. Not only does this commonly result in ammonia toxicity, but high levels of nitrogen are known to exacerbate disease by many pathogens such as Rhizoctonia, Pythium, Phytophthora, Fusarium, Armillaria,

Sclerotium, Pseudomonas, Corynebacterium, powdery mildews, rusts, cyst nematodes and many others.

Principle IV: Overwatering and under watering plants both exacerbate disease. Perhaps the most common plant problems, for most home owners and for ornamental plants in general, results from improper watering. Plants damaged by growing in water saturated soil can often be diagnosed by the foul hydrogen cyanide odor of the soil. Plants which are chronically drought stressed usually have leaves which have brown necrotic tips or edges. However, most people are not aware that overwatering or under watering can predispose plants to disease. Predisposition is the environmental modification of plant resistance making the plant more susceptible to disease. Exposing roots to saturated water conditions for as little as 18 hours damages the sensitive root membranes and results in leakage of nutrients from the root cells. This is called root exudation and many pathogens are attracted to root exudates and will preferentially attack roots which are leaking exudates. Overwatered plants often become more susceptible to root rot fungi such as Phytophthora and Pythium. Underwatering plants appears to make them less able to fight off invading pathogens and drought stressed plants are more susceptible to cankers caused by Botryosphaeria (Fusicoccum) and Hendersonula.

Principle V: The realistic way to manage plant disease is through an integrated management strategy that includes: 1) cultural practices, 2) epidemiology, 3) resistant varieties, 4) chemical pesticides, and 5) biological control.

Plant diseases are often extremely difficult to control and require a flexible approach which utilizes all of the tools available to us. In most situations, utilizing several methods to control disease is more practical and efficient than relying on a single method. An example of cultural control is mulching which will often reduce root rot caused by Phytophthora. Epidemiology is the study of factors affecting the outbreak and spread of infectious disease. Familiarity with the epidemiology of a pathogen will show how factors such as temperature, rainfall, wind and leaf wetness affect its ability to produce spores, spread and infect its host. Knowledge of epidemiology will often lead to the formulation of "models" which forecast or predict disease. Simple models such as rainfall for longer than 24 hours when the temperature is above 24 °C requires a fungicide spray are often the best and most useful means of both controlling a disease and reducing the amount of fungicide applied. Resistant varieties are often the ultimate way to control disease. However, the resistance may break down and for many ornamental plants resistance is simply not available. Chemical pesticides are often the only effective way to control disease. New IPM (Integrated Pest Management) guidelines stress using only enough pesticide to control the disease and to constantly look for ways to reduce pesticide applications. Biological control of plant pathogens is not as advanced as it is in insect control. While there are substantial numbers of biological control agents on the market, most are not very efficacious for control of diseases in the field.

Principle VI: A major approach to disease control is inoculum reduction. Inoculum is the pathogen or its parts which can cause plant infection. Reduction of inoculum is a two pronged approach, which includes inhibiting the pathogen and reducing its spread. Inoculum can be fungal spores or survival structures, bacteria, or virus particles, but it can also be plant debris, infested soil, infected roots, diseased parts of perennial plants, insect vectors or infected fruit or vegetative organs. Inoculum must be reduced or eliminated to control disease. Inoculum can be moved by wind, wind-blown rain, water, insects, seeds, infected transplants, animals, humans, and equipment and pruning tools. Care to prevent movement of inoculum to the plant will effectively reduce the inoculum and control disease.

Principle VII: Wood decay organisms result in wind damage and breakage in trees. While this principle seems self-evident, it is amazing how many homeowners are more interested in saving their tree than in the potential danger to their house, car or their own person. Wood decay organisms often damage only the dead, central heartwood portion of the tree causing the tree to be hollow. The tree shows very little outward signs of the damage since one or two outer rings of the trunk can conduct all the water and nutrients to sustain the tree. Fruiting bodies of the wood decay fungi, which resemble brackets or shelves, are signs that the tree may in danger of breaking and falling over. Two of our most common wood decay fungi in Southern California are Ganoderma and Laetiporus.

Principle VIII: Epidemics of introduced (invasive) species are more severe than epidemics of endemic species. New pests and diseases are constantly being introduced into Southern California simply because there are more people traveling today than ever before. Notices in airports and borders plead with visitors not to bring in plant material which may harbor new diseases. However despite our best efforts new diseases are introduced and because our plants have little resistance to introduced species and there are no competitors or predators to keep their numbers low, and a epidemic occurs which is very difficult to control. The cost to the US for introduced species is in excess of $137 billion. Because of the costs due to the loss of chestnut trees due to the introduction of chestnut blight, the loss of white pine due to the introduction of white pine blister rust, the loss of elm trees due to the introduction of Dutch elm disease and avocado losses due to the introduction of avocado root rot. Invasive diseases of these four trees alone have resulted in billions in losses at today's prices. Despite the massive losses due to introduced species our state and federal government continues to reduce quarantine efforts at our borders.

Principle IX: Quarantine is often the best method for combating disease. This principle is really a corollary of Principle VIII.

Principle X: Do no harm. This is really the first rule of plant pathology. Certainly our state and federal governments are ignoring this rule when they downgrade our quarantine system. However, this rule is really aimed at the farmer or homeowner who tries to control disease without really understanding plant pathology. Diseases are commonly exacerbated by overwatering and over-fertilization by well-meaning gardeners. Plant pathologists term efforts to control disease which result in worse disease as boomerang effects. A classic example of the boomerang effect occurs if a soil-borne pathogen reinvades fumigated soil. Without natural enemies to reduce its spread it often becomes far worse than it would have been before fumigation. Disease trading is another term which results when one disease is controlled but another is exacerbated by the same treatment.

## References


- What-is-plant-pathology, education: phytopath.ca, Retrieved 20, March 2020
- Plant-disease, science: britannica.com, Retrieved 12, June 2020
- Environmental-Horticulture, landscapes, pathology: ucanr.edu, Retrieved 07, January 2020

# Chapter 2

# Understanding Plant Pathogens

There are a wide variety of pathogens that cause plant diseases. These include the various plant pathogenic bacteria, fungi, viruses, etc. An understanding of these causal agents is critical to the protection and restoration of afflicted plants. This chapter addresses each topic in a detailed manner for a thorough understanding of the subject.

Pathogens include fungi, bacteria, nematodes, and viruses, all biological organisms that can cause disease symptoms and significantly reduce the productivity, quality, and even cause the death of plants. Pathogens can also infect agricultural animals.

Pathogens can be introduced and spread to host plants in many ways. Bacterial and fungal spores can be transferred by wind, in rain, and from soil via rain splashing onto plant tissues. Insects can vector or infect a plant with a pathogen when they feed on an infected host plant, and then move and feed on an uninfected plant. Pathogens can also spread through infected seeds, transplants, or contaminated equipment, irrigation water, and humans.

The phyto-pathogens are two types: biotic factors, which include all microbes and parasitic plants, and abiotic factors, which include all environmental factors. Essentially, the plant pathology is correlated with other sciences such as entomology, bacteriology, mycology, virology, and weed science due to deleterious effects of insects, bacteria, fungi, viruses/viroid, and weeds on plants, respectively. The first step of plant disease treatment is observation of definite and clear symptoms on the plants. These symptoms give an initial indication for the type and cause of plant disease, which may end with the death. The modern approach of plant disease control depends on biological control agents such as the production of antimicrobial agents and the production of genetic-improved strains of plants, which are more resistant to plant diseases. This approach is more favorable because it is friendlier with the environment and healthier for humans and animals. The infected part of the plant gives an indication of the type of plant disease, such as infected root which is usually correlated with root-rot disease. The plant diseases can be classified according to several parameters: disease symptoms, infected organ, infected plant type, and the type of phyto-pathogen. The latter is considered the more useful criterion used for plant disease classification, because it easily determines the disease cause, potential disease complications, and possible control methods. According to this criterion, plant diseases are classified into two types: infectious (biotic) diseases, which are caused by eukaryotes, prokaryotes, parasitic higher plants, viruses/viroid, nematodes, and protozoa, and noninfectious (abiotic) diseases, which are caused by different extreme environmental conditions.

### Infectious Diseases

There are wide range of phyto-pathogens which cause infectious plant diseases such as fungi, bacteria, viruses, viroids, mollicutes, parasitic higher plants, and protozoa. The infectious disease means the ability of phytopathogen to transfer from the infected plant to another healthy one and causes the same disease and the same symptoms. The most phytopathogens can inhabit the internal environment of plants; however, some others can live on the plant surface such as some fungi, bacteria, and parasitic higher plants.

## Diseases Caused by Parasitic Higher Plants

Some plant diseases are developed due to growing certain plants attached on or in other plants, where they take all required nutrients without benefit sharing; these plants are called parasitic higher plants. This abnormal relationship leads to weakness of healthy or host plant. The parasitic higher plants are usually found attached with the surface of the host plant, such as dodder, mistletoe, witch weed, and broomrape.

## Diseases Caused by Nematodes

The nematodes are one of most common phyto-pathogens which have definite symptoms. These symptoms only appeared in the infected site. The nematode infections in or on plants are widely distributed especially in proper environments such as moderate temperature and high humidity.

## Diseases Caused by Fungi

Interestingly, there are two main types of fungi appearing on plants: pathogenic and saprophytic. The pathogenic fungi live in or on plant tissues and cause serious complications for the vital physiological functions of plants, while saprophytic ones live in or on dead tissues. Accordingly, the diagnosis of plant disease must be exactly carried out. The exact diagnosis and determination of fungi take place by microscopical examination to identify the mycelial morphological characteristics, whatever fruiting structures and spores. After complete identification for the fungus and the symptoms of plant disease, the latter should be compared. Although microscopical examination is an essential and effective method for fungal identification, it only sometimes cannot lead to exact identification due to the absence of fungal fruiting structures and spores on infected plant tissue. Therefore, an alternative method must be used, such as using selective media for isolation, identification, or promotion of sporulation. On the other hand, some fungi need to be incubated under certain temperature, aeration, or light conditions to produce spores.

## Diseases Caused by Bacteria and Mollicutes

The appearance of bacterial growth in or on plant tissues means that bacterial plant disease may be present, because saprophytes may be present. Therefore, accurate bacterial identification must be carried out by using microscopical examination and physiological parameter determination. The selective media are essentially used in the bacterial identification to determine the bacterial genus and species in some cases. Moreover, the confirmatory test of bacterial pathogenicity may be carried out by inoculation of single pure bacterial colony in the healthy plant, reproducing the same symptoms that appeared on the infected one. Moreover, immunodiagnostic techniques or serodiagnostic assays can be used, such as agglutination and precipitation, fluorescent antibody staining, and enzyme-linked immunosorbent assay (ELISA). There are several advantages for these techniques such as quite sensitivity, fairly specificity, rapid, easy to perform, and it is expected that standardized, reliable antisera will be available soon. Furthermore, there are recent methods used for bacterial identification, which depend on the automated analysis of bacterial fatty acid profile. The molecular biological techniques are also widely used. There are uncommon microorganisms called mollicutes. These microorganisms are very small where they must be examined by an electron microscope. Mollicutes have polymorphism and lack cell wall–like mycoplasma. These microorganisms habit the young phloem cells as a convenient host, and cause severe plant diseases such as plant stunting, yellowing or reddening of leaves, proliferation of shoots and roots, production of abnormal flowers, and eventual decline and death of the plant. Mollicutes cannot be cultured on nutrient media except for the genus Spiroplasma. Mollicutes can be diagnosed by several parameters, such as symptoms determination, grafting, transformation, microscopical examination, susceptibility to tetracyclines, and so on.

## Diseases Caused by Viruses and Viroids

There are distinctive types of plant diseases caused by viruses/viroids. These diseases have definite and clear symptoms, which easily support disease diagnosis and are considered main advantage. Apart from this advantage, some recent techniques are widely used for disease diagnosis and virus identification, such as virus transmission tests to specific host plants by sap inoculation, grafting, certain insect, nematode, fungus, and mite vectors. Moreover, sero-diagnostic tests are used for this purpose such as enzyme-linked immune-sorbent assays, gel diffusion tests, micro-precipitin tests, and fluorescent antibody staining. The electron microscopy techniques as negative staining of virus particles in leaf dip or purified preparations are also used, as well as immune-specific electron microscopy. On the other hand, there are more accurate techniques used for disease diagnosis and virus/viroid identification, such as electrophoretic tests and hybridization of commercially available radioactive DNA complementary to a certain virus DNA or RNA, or viroid RNA, with the DNA or RNA present in plant sap and attached to a membrane filter (immunoblot).

## Diseases Caused by more than One Pathogen

Sometimes, some plants are exposed to co-infection by two or more pathogens, which lead to the same or different disease symptoms. Therefore, the differentiation and identification of these pathogens are very essential to exactly determine the disease cause.

## Noninfectious Diseases

Occasionally, some plant diseases have abiotic origin such as environmental factors; these diseases are called noninfectious diseases. Abiotic environmental factors have deleterious effects on plants under extreme conditions, because they can negatively effect on the vital physiological functions and may lead to death, for example, the presence of considerable amounts of toxics in the soil or in the air, deficiency of water, oxygen, or minerals, and extreme conditions for temperature, humidity, oxygen, CO, or light.

## Parasitism and Pathogenicity

The term parasitism called on the state in which an organism (parasite) lives on or in another one (host) to obtain its required nutrition. Usually, the parasitism is correlated with pathogenicity, which means the ability of an organism to cause a disease. However, the parasitism in some cases leads to a benefit relationship called symbiosis, in which both plant and organism alternate the benefits, such as bacterial nodules in the roots of legume plants and the mycorrhizal infection of feeder roots of most flowering plants. In the case of parasitism-pathogenicity relationship, the plant is diseased with the appearance of different symptoms such as increased respiration, disintegration or collapse of cells, wilting, abscission, abnormal cell division and enlargement, and degeneration of specific components such as chlorophyll. The most common plant pathogens are fungi, bacteria, mollicutes, parasitic higher plants, parasitic green algae, nematodes, protozoa, viruses, and viroids. These parasites cause serious plant diseases, because they have the ability to penetrate the plant tissues to feed and proliferate in it, and withstand the conditions in which the host lives.

These pathogens are also called obligate parasites because they can only live in their living hosts. On the other hand, there are certain pathogens such as most fungi and bacteria can live on either living or dead hosts and on various nutrient media, so they are called nonobligatory parasites. Some nonobligatory parasites can grow saprophytically on dead organic matter, and therefore called semi-biotrophs/facultative saprophytes. There is a type of a life called facultative parasitism, in which an organism grows saprophytically (necrotrophs); however, under certain conditions, they attack living plants and cause a disease; these parasites are called facultative parasites. The type or degree of parasitism does not affect the disease severity. For instance, many diseases caused by weakly parasitic pathogens are much more damaging to a plant than others caused even by obligate parasites. Lysozymes are a main mechanism of most nonobligatory parasites by which they can degrade the plant cell wall and subsequently cause invasion and infection.

### Host Range of Pathogens

Phytopathogens differ among each other with respect to the plant type, the location of infection, and the age of the organ or tissue (location of infection). The specificity of plant pathogens has various degrees; some pathogens have only one target species of plant, while other pathogens can attack only one genus of plants, and eventually some others have a wide range of hosts, belonging to many families of higher plants. Phytopathogens differ among each other with respect to the location of infection; some of them grow on roots, stems, leaves, fruits or vegetables, and phloem or xylem. Some phytopathogens can only infect the seedlings or the young parts of plants, while the others can only infect the mature tissues.

## Plant Pathogenic Bacteria

Bacteria are single-celled microorganisms, generally ranging from 1-2 μm in size that cannot be seen with the unaided eye. Plant associated bacteria may be beneficial or detrimental. All plant surfaces have microbes on them (termed epiphytes), and some microbes live inside plants (termed endophytes). Some are residents and some are transient. Bacteria are among the microbes that successively colonize plants as they mature. Individual bacterial cells cannot be seen without the use of a microscope; however, large populations of bacteria become visible as aggregates in liquid, as biofilms on plants, as viscous suspensions plugging plant vessels, or colonies on petri dishes in the laboratory. For beneficial purposes or as pathogens, populations of 106 CFU (colony-forming units/milliliter) or higher are normally required for bacteria to function as biological control agents or cause infectious disease.

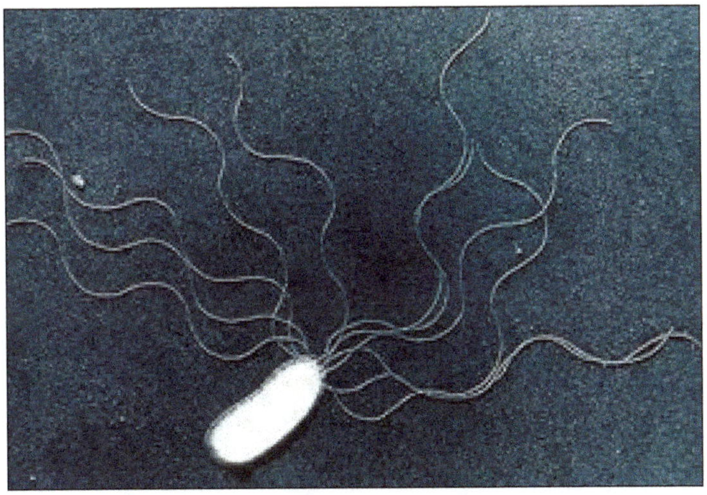

Plant pathogenic bacteria cause many serious diseases of plants throughout the world, but fewer than fungi or viruses, and they cause relatively less damage and economic cost. Most plants, both economic and wild, have innate immunity or resistance to many pathogens. However, many plants can harbor plant pathogens without symptom development (asymptomatic).

Most phytopathogenic bacteria are rod shaped bacillus the only exception being Streptomyces (family Actinomycetes) which is a filamentous (thread-like, filiform) bacteria. Also, most of these bacteria have flagella which are whiplike structures projecting from a bacterium that functions as an organ of locomotion. Some species of bacteria have only one flagellum (monotrichous) or a tuft of two or more flagella at one end of the cell. These are called polar flagella. Other species will have flagella distributed over the entire surface of the cell. These are termed peritrichous flagella.

Of the over 15,000 identified species of bacteria most are saprophytic and are of great benefit in decomposing dead and rotting organisms thereby releasing their nutrients back into the environment. This is the most important roll that bacteria play in nature. Plants rely on nitrogen from the soil but cannot directly acquire it from the gaseous nitrogen in the atmosphere. The primary way nitrogen is supplied to plants is through the mineralization of organic material in the soil. However, nitrogen fixation by bacteria such as Rhizobium spp. and Cyanobacteria spp. is almost as important as mineralization, and is a primary source of nitrogen. As these bacteria metabolize they convert gaseous nitrogen into nitrates or nitrites that become available to plants.

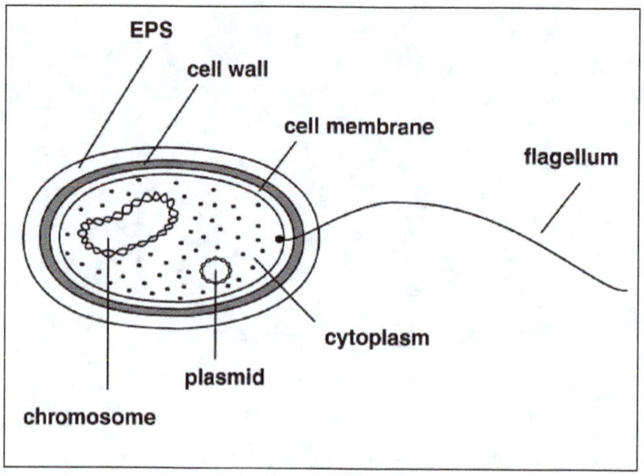


Most phytopathogenic bacteria are aerobic (live in the presence of oxygen) and some are facultative anaerobes which can grow with or without oxygen. Some bacteria have thick, rigid cell walls which will retain dye from a cell staining method developed by Christian Gram, while other bacteria will not accept this stain. This method of staining results in the bacteria being classed as Gram-positive or Gram-negative and is an important factor in identification and classification. Gram-positive bacteria appear purple and Gram-negative bacteria appear pink under magnification. Bacteria are also distinguished by the different kinds of enzymes they either can or cannot use for nourishment and the nutrient media on which they can grow.

Rod shaped bacteria reproduce asexually by the process of binary fission (the transverse splitting in two of a bacterial cell). This process takes place when the cytoplasmic membrane grows inward dividing the cytoplasm into two approximately equal parts. When the cell walls are completely formed the cell splits into two cells. During this process the nuclear material duplicates itself and becomes distributed equally between the two cells. Bacteria can reproduce at a very rapid rate; some species can divide every 20 minutes under ideal conditions. It is conceivable that a single bacterium could produce one million progeny in less than 24 hours. However, with limited food supply, environmental conditions and other factors the optimum conditions rarely occur in nature.

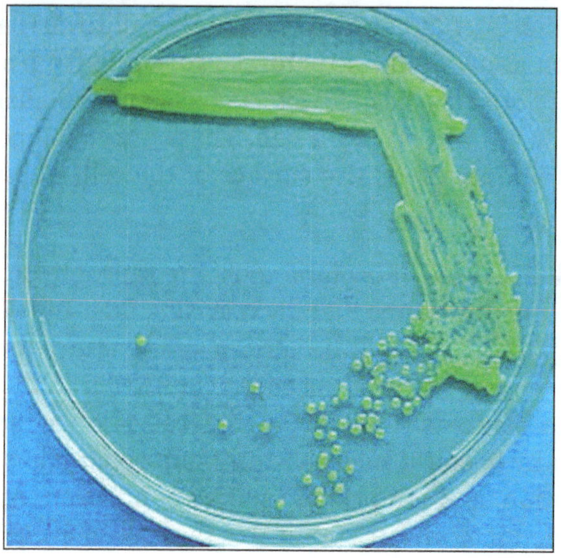

There are around 200 species of phytopathogenic bacteria and almost all of them are parasites within the plant, on its surface, in plant debris or in the soil as saprophytes. Dissemination of bacteria can be accomplished by several means. Some bacteria can survive on inanimate objects, in water or inside insects. It is important to know the survival characteristics of bacteria for effective management strategy and intervention in dissemination. Some species have the ability to move short distances in water on their own power by use of their flagella. Most bacteria, however, are disseminated by passive agents such as air and insects, water and soil movement, and to a lesser degree by humans, water and other animals. Infected seeds and transplants can also be a source of inoculums. Most bacteria require a wound or natural opening (e.g. stomata, lenticels or hydathodes) to gain entry into the host tissue and also require warm, moist conditions to establish a colony. Windblown soil and sand will commonly cause wounds which can facilitate bacterial infections.

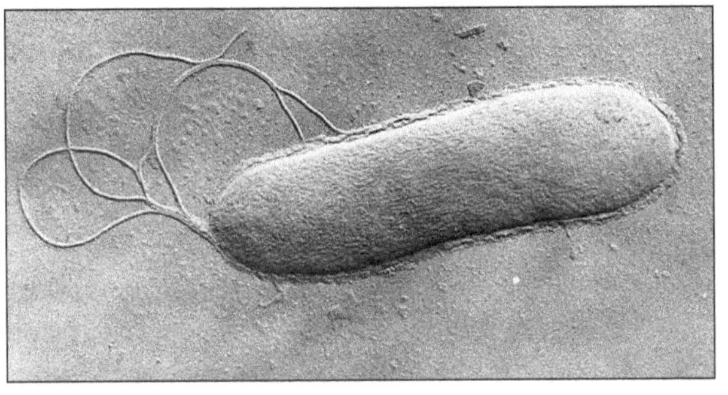

Bacteria colonize a host by growing between the cells and absorbing the cells nutrients that leak into intercellular space or grow within the vascular tissue of the plant. Depending on the species of bacteria and the tissue infected they produce and release enzymes that degrade cell walls, growth regulators that alter the plants normal growth, toxins that degrade cell membranes and complex sugars that plug water conducting tissue.

## Pseudomonas Syringae

Pseudomonas syringae is a phytopathogenic bacterium that causes diseases of monocots, herbaceous dicots, and woody dicots, worldwide. On woody plants, reports of disease due to P. syringae have markedly increased in the last years and the diseases have been recognized as a major threat to the primary products of agroforestry practices. Detection in Italy of a new highly aggressive population of P. syringae in 2008 on kiwifruit, which caused severe epidemics in the following years throughout the kiwifruit-growing areas of Asia, Europe, Oceania, and South America, rendered the entire kiwifruit industry vulnerable to the disease. Similarly, occurrence of an aggressive population of P. syringae on horse chestnut in 2002 in the Netherlands has rapidly established itself as a major threat to horse chestnut throughout Northwest Europe. To better understand the origin of such disease epidemics, a thorough knowledge of the pathogen is needed in sensu lato.

## Agrobacterium Tumefaciens

Agrobacterium tumefaciens causes crown gall disease on a wide range of host species by transferring and integrating a part of its own DNA, the T-DNA, into the plant. This unique mode of action has also made the bacterium an important tool in plant breeding. After attachment of Agrobacterium to plant cells and expression of multiple virulence (vir) genes, several effector proteins, together with T-DNA, are transported into the plant cell by a type-IV-secretion system. Plant factors assist with T-DNA integration into the plant. After integration, expression of the T-DNA-encoded oncogenes iaaH, iaaM, and ipt induces biosynthesis of auxin and cytokinin. Increased levels of these phytohormones result in enhanced proliferation and formation of crown galls. Despite the transfer of bacterial proteins into the plant cell, most Agrobacterium strains do not elicit a hypersensitive response (HR), which is associated with rapid and localized death of cells. Such a response often occurs when plants are challenged by bacterial pathogens and serves to restrict the growth and spread of pathogens to other parts of the plant. Accordingly, no systemic, broad-spectrum resistance response throughout the plant (systemic acquired resistance, SAR) is induced. Within the first several hours of co-cultivation, pathogen defense response pathways are activated more or less strongly depending on the plant system and Agrobacterium genotype used for infection. Defense responses become stronger during crown gall development. Furthermore, the physiological behavior of the transformed cells changes drastically.

## Host Responses to Agrobacterium Tumefaciens Before T-DNA Transfer

### Pathogen Defense

The recognition of microbial pathogens plays a central role in the induction of active defense responses in plants. The conserved flagellin peptide flg22 is recognized by the receptor kinase FLS2 and induces the expression of numerous defense-related genes to trigger resistance to pathogenic bacteria. However, the genus Agrobacterium fails to induce this type of rapid and general defense response because of an exceptional divergence in the N-terminal conserved domain of flagellin. When comparing early gene expression changes after infection with the virulent Agrobacterium strain C58 with application of the bacterial peptide elf26 (after 1 and 3 h, respectively), dampening of host responses becomes apparent with Agrobacterium treatment. The elf26 peptide, a highly conserved motif of one of the most abundant proteins in microbes recognized by the receptor kinase EFR, is a fragment of the elongation factor Tu (EF-Tu). EF-Tu triggers innate immunity responses associated with disease resistance in Arabidopsis. While treatment with pure elf26 induces gene expression changes of 948 Arabidopsis genes, only 35 genes are induced after infection with the virulent Agrobacterium strain C58, suggesting that the bacterium somehow neutralizes the response to elf26 by the host plant. It should be mentioned that the Arabidopsis ecotype and age (seedling vs. adult stalk) used in the studies may also account for some of the differences in defense response.

Concerning the transcriptional activation of genes involved in early plant defense responses, several studies have come to different conclusions. Ageratum conyzoides cell cultures showed differential expression of defense genes as early as 24 h post infection with a non-oncogenic hyper-virulent Agrobacterium strain. In tobacco suspension cultures infected with different Agrobacterium strains, transcription of defense genes increased within 3–6 h, but started to decrease with the onset of T-DNA-transfer. A study using suspension-cultured cells of Arabidopsis did not show changes in transcript levels within 4 to 24 h but activation of defense genes 48 h after. When agrobacteria are inoculated at the base of wounded Arabidopsis stems just very few defense genes are activated 3 h post infection compared to uninfected wounded stems. In contrast to cell cultures, the latter experimental setup does neither require phytohormone pre-treatment nor virulence gene induction prior to infection. Phytohormone pre-treatment of the cell culture systems of the earlier studies may alter host cell defense responses. Thus, discrepancies between these studies probably result from the different plant inoculation systems used. Nevertheless, agrobacteria can abuse host defense responses for T-DNA delivery. The mitogen-activated protein kinase MPK3 phosphorylates the Arabidopsis VIP1 protein, inducing VIP1 relocalization from the cytoplasm to the nucleus. Nuclear localization of VIP1 increases T-DNA transfer and transformation efficiency.

## Phytohormones

Agrobacteria produce auxin and cytokines themselves in order to modulate plant. These phytohormones have been determined in the cells as well as cultivation medium. It was postulated that biosynthesis of the phytohormones is catalyzed by enzymes of the T-DNA encoded oncogenes, as transcripts and proteins of these genes were detected in agrobacterial cells. Pronounced amounts of auxin have been determined in the virulent Agrobacterium strain C58 and at lower levels also in plasmidless and T-DNA depleted strains. More recent data have confirmed the latter results. The finding that a strain without a Ti-plasmid still can make auxin implies localization of genes also outside of the Ti-plasmid. However, this assumption is not supported by sequencing data for strain C58. Genes known to be involved in auxin biosynthesis seem to be encoded only by the T-DNA of the Ti-plasmid. The presence of iaaH and iaaM transcripts by PCR in Agrobacterium cells of strain C58 and confirmed the earlier findings. It remains to be proven whether these genes are responsible for auxin production or if auxin is synthesized by a different mechanism in Agrobacterium cells. The mechanism for cytokinin biosynthesis by agrobacteria is far better understood. In nopaline utilizing Agrobacterium strains cytokinin is produced in high amounts by the Ti-plasmid encoded trans-zeatin synthesizing (tzs) enzyme of which the gene is located in the vir regulon. A substantial smaller source for cytokinin production is isopentenylated transfer RNA (tRNA) catalyzed by the chromosomal-encoded enzyme tRNA: isopentenyltransferase (MiaA) present in all Agrobacterium strains.

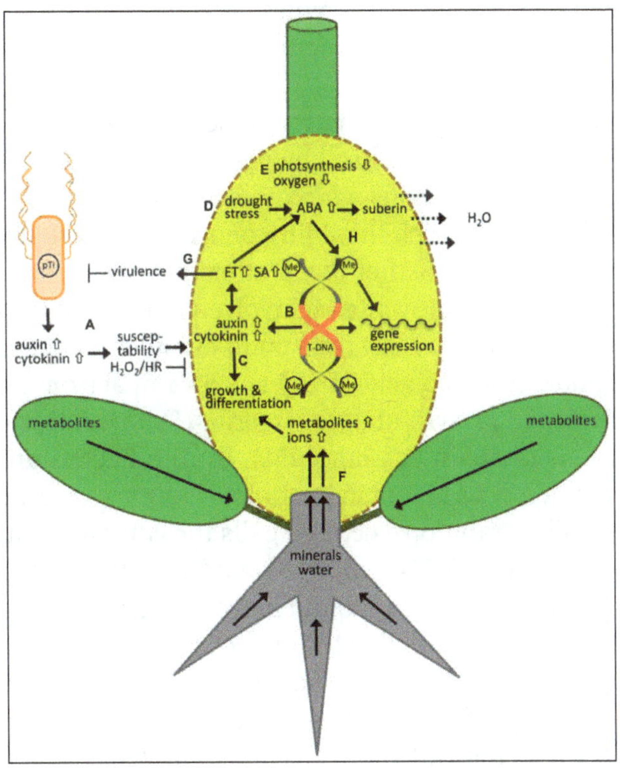

Responses of the model plant Arabidopsis thaliana to Agrobacterium tumefaciens and crown gall development. (A) Virulent (pTi) agrobacteria cells themselves produce and release cytokinin and auxin, which increase host susceptibility and inhibit hydrogen peroxide production ($H_2O_2$) and hypersensitive response (HR) at initiation of infection. (B) After integration of the bacterial T-DNA into the plant genome, cytokinin and auxin is synthesized by T-DNA encoded enzymes and accumulate inside the tumor. (C) This causes massive changes in the gene expression pattern, resulting in metabolomic and morphological adaptations that are necessary for tumor growth and differentiation. (D) Loss of water is minimized by drought stress protecting mechanism, which causes an increase in the levels of the stress hormone ABA, and ABA-dependent suberization of cells to prevent water loss. Evaporation of water ($H_2O$) from the disrupted crown gall surface drives the flow of water and minerals into crown galls. (E) Because photosynthesis is down-regulate the oxygen levels are low, the tumor produces C and N compounds heterotrophically and gains energy mainly anaerobically by alcoholic fermentation. (F) Consequently the developing tumor becomes a metabolic sink for the host plant, which accumulates metabolites produced by source leaves and minerals taken up by the roots. (G) Auxin and cytokinin also cause an increase in ethylen (ET) which together with salicylic acid (SA) inhibits agrobacterial virulence. (H) ABA also induces DNA methylation of the plant genome, thereby regulating gene expression of drought-stress responsive genes. Overall, the crown gall genome becomes hypermethylated (Me) after Agrobacterium infection and possibly contributes to the strong changes in gene expression during tumor growth. The oncogenes of the T-DNA remain unaffected by methylation of the plant genome.

Earlier studies have shown that pre-treatment of explants with either auxin alone or both auxin and cytokinin increase T-DNA transfer efficiency and stable transformation as well as crown gall growth. In this respect, Agrobacterium produced phytohormones play a role at very early time points of infection, before T-DNA-encoded enzymes catalyze synthesis of cytokinin and auxin in the transformed host cell. Concerning the mechanism causing an increase in susceptibility it was speculated that phytohormones induce plant cell division and that the cell cycle phase influences agrobacterial attachment and stable transformation. It seems likely that phytohormone-mediated modification of the physiological state of the cell increases competence for T-DNA transformation and integration. More recent investigations addressed the question about the molecular mechanism and the signaling pathways by which these phytohormones influence host cell susceptibility. Transcriptome microarray data from 3 h after inoculation of Agrobacterium strain C58 into Arabidopsis stems revealed that the genes known to be involved in phytohormone-dependent signaling are not induced in host cells at this very early time point of infection before transfer of the T-DNA. It has been shown that indole-3-acetic acid (IAA) has an impact on agrobacterial virulence by inhibiting vir gene induction and growth of agrobacteria. However, this effect was observed with relatively high concentrations of auxin (25–250 μM). In Agrobacterium cells the total (free and conjugated) IAA content is 0.3 ± 0.1 μM and in Arabidopsis stems 3 h after inoculation with strain C58 it is 2.1 ± 1 μM, whereas in Arabidopsis crown galls the content is ca. 10 times higher (17.3 ± 8.8 μM) due to the expression of the T-DNA encoded iaaH and iaaM genes and their enzyme activity. Application of 1 μM IAA, a concentration found in wounded and uninfected Arabidopsis stems (0.8 ± 0.2 μM), stimulated growth of Agrobacterium cells, whereas growth stimulation vanished at 10 μM and higher IAA concentrations. It is known that the effect of auxin is strongly dose dependent with a growth promoting effect at low concentrations and an inhibitory effect at high concentrations, which slightly varies dependent on the plant and tissue type. One may speculate that at initiation of infection, the relatively low auxin levels of agrobacterial cells and/or of wounded plant tissue stimulate growth of agrobacteria, whereas the higher concentrations produced in the crown gall inhibit virulence as well as growth of Agrobacterium. Such an antagonistic auxin effect would promote transformation of the host cell at the beginning of the infection process and inhibit agrobacterial virulence and growth to prevent further transformation events in developing crown galls. In contrast to auxin, the role of cytokinin signaling in plant susceptibility is well known. Recently, it has been shown that cytokinin secreted by Agrobacterium controls virulence via bacterial cell growth and vir gene expression at early stages of the infection process. Some, but not all plant species showed a cytokinin-dependent increase in transformation efficiency. Agrobacterium-derived cytokinin not only acts on bacterial physiology but also influences host gene expression via the classical cytokinin-dependent signaling pathway including cytokinin receptors and the phosphotransfer cascade. Activation of this signaling cascade through agrobacterial-derived cytokinin results in inhibition of gene expression of the Arabidopsis MYB family transcription factor, MTF1. MTF1 turned out to be a negative regulator of transformation susceptibility by blocking expression of the integrin-like protein At14a, a plant membrane receptor. A serves as anchor points for bacterial attachment at the host cell surface. Thus, at early stages of infection agro-bacterial auxin and cytokinin manipulates plant phytohormone signaling pathways to prepare the host cell for transformation.

In addition to auxin and cytokinin, plant defense signaling involves a network of interconnected pathways in which salicylic acid (SA) and jasmonic acid (JA) together with ethylene (ET) function as essential signaling molecules. Exogenous application of the plant defense molecule SA to Agrobacterium cells inhibited expression of vir genes including tzs, bacterial growth, bacterial attachment to plant cells and virulence. However, at initiation of infection (3 h post infection) neither SA nor JA levels nor the genes of these signaling pathways are elevated in Agrobacterium-infected Arabidopsis tissues. At this time point only the level of 1-amino-cyclopropane-1-carboxylic acid (ACC), an ET precursor, is increased in the presence of both virulent and disarmed Agrobacterium strains, but not expression of marker genes of the ET-dependent defense-signaling pathway. Inoculation of melon (Cucumis melo) explants with Agrobacterium also increases ET production. ET is known to trigger plant auxin production due to increased expression of plant genes involved in auxin biosynthesis. Auxin enhances host susceptibility whereas plant ET production has a negative effect on agrobacterial virulence. Application of ACC reduces Agrobacterium-mediated gene transfer to melon explants whereas addition of aminoethoxyvinylglycin, an inhibitor of ACC synthase, increased it. A reduction in transformation efficiency results from suppression of vir gene expression, but not Agrobacterium growth. The promoting effect of low auxin concentrations on agrobacterial growth and the inhibiting effect of ET on virulence illustrates that both, Agrobacterium and the host plant control host cell transformation. Taken together, at early stages of the infection process, cytokinin and auxin produced by Agrobacterium cells have a promoting effect on transformation efficiency, which is in part counteracted by the inhibitory effect of host plant-derived ET and SA on agrobacterial virulence. Thus, the correct phytohormone balance decides on the success of infection.

## Hypersensitive Response

Examination of early events in pathogenesis has demonstrated that virulent Agrobacterium does not induce HR in Arabidopsis. Moreover, Agrobacterium is able to suppress HR induced by Pseudomonas syringae pv. phaseolus in plants. This suppression is dependent on the activity of the iaaH and iaaM oncogenes which encode enzymes for auxin synthesis, since several Agrobacterium transposon mutants in the iaa genes failed to inhibit a HR. Likewise, transcription of several genes involved in oxidative stress signaling are only induced by the oncogenic, but not the T-DNA-depleted Agrobacterium strain. Production of $H_2O_2$ precedes HR, which is degraded via a chromosomally encoded catalase of Agrobacterium. $H_2O_2$ acts both as a local trigger for the programmed cell death and as a diffusible signal for the induction of cellular protectant genes in surrounding cells. Apart from its signaling functions, $H_2O_2$ is also involved in toughening of cell walls in the initial stages of plant defense by cross-linking of cell wall structural proteins. Accumulation of $H_2O_2$ is prevented only at the early stages of agro-bacterial infection, but proceeds in the course of tumor development.

## Host Responses to Crown Gall Development

### Morphological Adaptations

Development of crown galls is accompanied by profound changes in the gene expression profile, metabolism, and morphology. The uncontrolled synthesis of auxin and cytokinin by cells transformed with a T-DNA of tumorigenic Ti-plasmids drives tumor development, while the auxin to cytokinin ratios determine the crown gall morphology. In the early days of studies about the molecular basis of crown gall development it was observed that mutations in the tmr locus encoding ipt cause rooty crown galls and those in the tms loci coding for iaaH and iaaM induce shooty phenotypes. A recent study on the T-DNA locus Atu6002 of strain C58 indicated that when the encoding protein C is expressed, it increases host cell sensitivity to auxin. In addition to the T-DNA-encoded genes, the expression of several host genes involved in auxin and cytokinin metabolism and signaling are expressed in crown galls. Cytokinin and auxin together with ET are known to be essential for growth of crown gall tumors and differentiation of cell types with different morphology and function. Particularly, ET has been shown to be essential for the formation of vascular tissue and crown gall tumor development. Application of the ET synthesis inhibitor aminoethoxyvinyl-glycine prevents vascularization in castor bean (Ricinus communis) stems and inhibits tumor growth completely. When the ET-insensitive tomato (Lycospersicon esculentum) mutant, never ripe, is infected with virulent Agrobacterium cells it does not develop tumors despite integration and expression of the T-DNA encoded oncogenes for auxin and cytokinin biosynthesis. Thus, neovascularization is a prerequisite for crown gall development.

Growth and expansion of crown gall tumors cause disruption of the epidermal cell layer and thereby loss of guard cells and an intact cuticle. Accordingly, expression of genes involved in cutin biosynthesis is down regulated. As a disrupted surface area provides access for pathogens and leads to uncontrolled loss of water for the host plant, the crown gall surface has to be sealed. This is achieved by differentiating a periderm-like surface layer. The polymerization of sobering monomers involves peroxidases for which $H_2O_2$ is the electron donor. Thus, $H_2O_2$ produced in crown galls functions in strengthening of cell walls rather than in induction of a HR. The stimulus for inducing suberization is drought stress-mediated ABA signaling. Drought stress signaling seems to play a central role in crown gall development. ABA accumulates in crown galls in high amounts and transcription of a set of drought and/or ABA-inducible genes is elevated. ABA synthesis is triggered by ET as demonstrated by the application of various inhibitors of ET or ABA biosynthesis and the use of ET-insensitive or ABA-deficient tomato mutants. Among the genes which play a role in drought stress protection of crown gall tumors is *FAD3*, encoding a fatty acid desaturase. The fad3-2 mutant with impaired biosynthesis in α-linolenic acid (C18:3) develops much smaller crown gall tumors particularly in low but not high relative humidity. Elevated levels of C18:3 were found in the phospholipid fraction of Arabidopsis crown gall tumors and maintain membrane integrity under drought stress conditions. In addition to gene expression changes, crown galls accumulate high amounts of osmoprotectants, such as proline (Pro), gamma amino-butyric acid (GABA), and alpha-aminoadipinic acid. The retarded tumor growth in abi and aba mutant plants underlines the importance of an ABA-mediated drought stress-signaling pathway in crown gall development.

## Nutrient Translocation and Metabolism

Expression profiles of genes involved in energy metabolism, such as photosynthesis, mitochondrial electron transport, and fermentation together with physiological data revealed that Arabidopsis tumors produce C and N compounds heterotrophically and gain energy mainly anaerobically by alcoholic fermentation. The change from autotrophy to heterotrophy reduces the oxygen level in crown gall tumors thereby inducing expression of hypoxia-sensitive genes, such as SAD6. This gene encodes a stearoyl-acyl carrier protein desaturase, which belongs to a class of enzymes known to catalyze the first step in fatty acid desaturation, an oxygen-dependent process. Despite limited oxygen availability in crown galls, SAD6 provides the monounsaturated fatty acid, oleic acid, for membrane phospholipids. Thus, expression of SAD6 maintains fatty acid desaturation under hypoxic conditions.

Crown gall tumors primarily use organic carbon and nitrogen for growth and are therefore a strong sink for the host plant. Metabolites and minerals have to be provided by the host plant and translocated into the crown gall tumor. The mechanisms of nutrient translocation and their accumulation have been studied on crown gall tumors by applying cytological staining, eletrophysiological, and $^{14}CO_2$ tracer techniques as well as a viral movement. Solutes enter the crown gall tumor via vascular tissue, which is connected to that of the host plant and consists of phloem for the transport of assimilates and xylem for water and minerals. Assimilates are produced by source leaves and are apoplastically and symplastically unloaded from the phloem in crown gall tumors. High apoplastic invertase activity indicated that sucrose is unloaded apoplastically. After cleavage of sucrose by sucrose-degrading enzymes, hexoses can be taken up via hexose transporters into tumor cells. Arabidopsis crown galls show elevated expression of several genes encoding sucrose degrading enzymes and a monosaccharide transporter. In addition, a high-affinity hexose transporter has been isolated from meristematic tobacco cells transformed with a tumor inducing T-DNA and was characterized as energy independent hexose uptake transporter. Application of the membrane impermeable fluorescent probe, carboxyfluorescein (CF) to source leaves and transient expression of the GFP-labeled potato virus X (PVX) coat protein (CP), exclusively exploiting plasmodesmata for distribution, demonstrated the existence of a symplastic transport pathway between the phloem and tumor cells. Both reporters show extensive cell-to-cell movement in the parenchyma of crown gall tumors but not in uninfected stem tissues of different plant species ranging from symplastic (Curcubita maxima) to apoplastic loaders (R. communis, Nicotiana benthamiana). The disrupted and enlarged surface of the crown gall tumor drives water and mineral translocation into crown gall tumors since the evaporation rate of crown galls exceeds that of leaves and non-infected stems. The periderm-like layer of suberized cells that covers the crown gall surface provides a considerable diffusion resistance against water vapor, but it is not an impermeable barrier for water. Cations and anions are taken up into the tumor cells through the function of membrane-localized channels and transporters expressed in the crown gall. Potassium channel mutants with impaired crown gall growth underline the importance of optimal nutrient supply for growth.

## Defense Responses

Gamma aminobutyric acid and Pro not only serve as osmoprotectants in drought-stress related processes of the host plant, but have also an impact on Agrobacterium virulence. GABA produced in crown gall tumors can be taken up by Agrobacterium cells and causes a delay in accumulation of 3-oxo-octanoylhomoserine lactone (OC8HSL) and Ti plasmid conjugation. GABA activates the AttKLM operon of which the AttM lactonase degrades the quorum sensing signal, OC8HSL, thereby turning on quorum quenching to protect the host plant against infections with bacterial pathogens. However, Pro interferes with the import of GABA and thereby prevents GABA-induced degradation of the bacterial quorum sensing signal OC8HSL. Thus, Pro antagonizes the GABA-induced degradation of OC8HSL and therefore may be used by the pathogen to by-pass the GABA-based host plant defense.

In addition to growth and developmental processes regulated by auxin and cytokinin, crown gall biology also involves pathogen defense signaling pathways. Hormones such as SA, JA, and ET are the primary signals inducing defense responses. In Arabidopsis crown galls the levels of SA and ET, but not JA, are elevated. JA has no obvious impact on crown gall tumor development, as the development on Arabidopsis JA-insensitive mutants is wild type-like. SA and ET contents together with the expression of pathogen-related marker genes of the SA- and ET-dependent signaling pathways increase with accumulation of the T-DNA-encoded iaa and ipt transcripts. Thus, auxin and/or cytokinin seem to be important for defense signaling in crown gall tumors, since the non-tumorigenic Agrobacterium strain which contains a disarmed pTiC58 does not induce expression of marker genes of the SA- and ET-dependent signaling pathways. It is known that high levels of auxin and cytokinin stimulate ET synthesis and its accumulation in crown galls. In contrast to ET, the classical marker genes of the SA-dependent signaling pathways are not induced most likely as a result of the high auxin content, which has been shown to inhibit SA responses to avoid the induction of SAR. Despite the lack of induction of SA-dependent defense signaling, Arabidopsis mutant plants with high SA levels strongly reduce while those with low SA levels promote tumor growth. Instead of inducing host defense pathways, high SA levels act directly on oncogenic agrobacteria by inhibiting vir gene expression and thereby reducing agrobacterial virulence. Besides SA-mediated inhibition of Agrobacterium virulence, SA activates the AttKLM operon, just like GABA does, to down regulate quorum sensing in Agrobacterium. Thus, activation of quorum quenching by auxin, SA, and GABA, is part of the plant defense program against Agrobacterium in the developing crown gall. In addition to SA, ET and IAA also inhibit the vir regulon and T-DNA transfer into plant cells. Thus, the interaction between the host plant and Agrobacterium is very much based on phytohormone cross talk which provides a balance between pathogen-defense by the host and crown gall development promoted by Agrobacterium.

## Xanthomonas Species

Scientific Name of Xanthomonas species are Xanthomonas axonopodis, Xanthomonas arboricola, Xanthomonas mcampestris and Xanthomonas hortorum. Xanthomonas species are gram negative, aerobic, rod-shaped bacteria with cells measuring 0.4 to 1.0 µm wide by 0.7 to 2.0 µm long. Colonies of Xanthomonas species usually are mucoid, convex, and yellow when grown on yeast dextrose calcium carbonate (YDC) media or sucrose peptone agar (SPA). All members of the genus Xanthomonas are catalase positive and oxidase negative. The classification of plant pathogenic Xanthomonas species has been a complicated affair, subject to numerous revisions over the years. At one time most strains were classified as pathovars within the single species Xanthomonas campestris, but since the revision by Vauterin, et al, the name X. campestris has been reserved for those strains affecting crucifers.

## Hosts, Signs and Symptoms

The numerous species and pathovars of Xanthomonas cause disease on a variety of ornamental plants. X. arboricola pv. poinsettiicola has been reported on three species of poinsettia and other species of the Euphorbiaceae including Euphorbia milii (crown-of-thorns) and Codiaeum variegatum (croton). X. axonopodis pv. begoniae causes bacterial blight of begonia. Bacterial leaf spot and blight of Araceae, including Dieffenbachia, Anthurium, Philodendron, and Syngonium is caused by X. axonopodis pv. dieffenbachiae. Black rot on ornamental kale is caused by X. campestris. The causal agent of bacterial leaf spot of English ivy and other Araliaceae species is X. hortorum pv. hederae. Bacterial blight of Pelargonium (geranium) is caused by X. hortorum pv. pelargonii. Other important ornamental hosts of Xanthomonas spp. include Ficus, Hydrangea, and Zinnia. For bacterial spot on ornamental Prunus species, refer to Xanthomonas arboricola pv. pruni. For bacterial spot on ornamental pepper, refer to bacterial spot of pepper and tomato. Signs of bacterial leaf spot can only be observed with a microscope. Infected leaf tissue will exhibit bacterial streaming when mounted with water on a slide and observed at 100x.

Symptoms usually involve necrotic, often water-soaked leaf spots, typically most severe on lower leaves. The spots may be very small at first but develop an angular shape as the movement of the bacteria is limited by leaf veins. Eventually several spots may enlarge, coalesce and cause death of mature leaves. Other symptoms can occur depending on the host plant. In some cases there is a chlorotic halo around the spot. V-shaped brown necrotic areas pointing in the direction of the petiole may develop on leaves of begonia and geranium. On poinsettia, symptoms of leaf spot first appear as water-soaked, gray pinpoint spots. As lesions develop they turn yellow to tan and become angular in shape and the necrotic areas may become surrounded by a yellow halo. On begonia, symptoms include water soaked lesions of the foliage, which are especially visible from the underside of the leaves. Symptoms on hydrangea first appear as water-soaked spots (1 to 4 mm in diameter), typically on lower leaves, which eventually darken to a reddish-purple.

X. axonopodis pv. begoniae, X. axonopodis pv. dieffenbachiae, and X. hortorum pv. pelargonii can cause systemic infection of their hosts. Symptoms of systemic invasion start with yellowing of the older leaves and petioles and wilting. Systemically infected leaves or flowers easily break off and may show dark brown streaks within the petioles.

Other genera of bacteria, especially Pseudomonas and Acidovorax, are frequently associated with bacterial leaf spots of ornamentals. Xanthomonas can be difficult to differentiate from leaf spots caused by other bacteria based on leaf symptoms.

## Ecology and Epidemiology

Species of Xanthomonas are easily spread in infected plant material such as seeds or cuttings, and by mechanical means such as contaminated pruning tools. Most often the bacteria are transferred to new plant tissues by splashing rain and irrigation water splash. The pathogen enters the plant through stomata, other natural openings, or wounds. Disease development is favored by warm, wet conditions which favor bacterium reproduction and dissemination. Bacterial leaf spot symptoms appear on host foliage within three to seven days after infection. This bacterium can also become systemic in some hosts. Xanthomonas species overwinter in diseased plant debris and some species have been documented existing as epiphytic populations on plant surfaces without causing symptoms.

## Geographic Distribution

Xanthomonas leaf spots and blights are ubiquitous worldwide, especially where conditions are warm and humid. Most species and pathovars are widely distributed but some are considered to be of quarantine importance. Both X. arboricola pv. poinsettiicola and X. axonopodis pv. dieffenbachiae are on the EPPO A2 List.

X. arboricola pv. poinsettiicola was originally reported in 1951 from India. In 1960, this pathovar was discovered on poinsettia in Florida. More recently X. arboricola pv. poinsettiicola has been found causing leaf spots in Taiwan, Zhejiang Province, China, Germany, and Italy. In the United States, reports of Xanthomonas leaf spot on poinsettia have mostly been restricted to Florida.

## Management

### Cultural

Starting with clean planting material (seed, cuttings, or plugs) is essential for preventing outbreaks of bacterial leaf spot. Management of this disease once it occurs requires strict sanitation. Plug trays with symptomatic plants should be discarded, and adjacent trays either discarded or placed in isolation, since they may have latent or epiphytic populations of the bacteria. Wetting foliage from overhead irrigation and overspray from lawn irrigation systems should be avoided, or at least timed to minimize the duration of leaf wetness. Adequate plant spacing is important to allow for air movement and more rapid drying of foliage. Management must include elimination of all stock plants with Xanthomonas leaf spot and disinfection of all surfaces in contact with diseased plants. Avoid handling plants when wet. Finish work in uncontaminated areas before starting work in areas that have had diseased plants, because bacteria can spread on hands and equipment.

### Chemical and Biological

Pesticides containing copper sulfate and copper octanoate (copper soap) are broadly labeled for control of leaf spots on ornamentals, but are only partially effective. These products should be applied preventatively in late spring or at first sign of disease to help limit disease spread. In order to slow the development of copper-insensitive strains, these products should be rotated with either the biological Bacillus subtilis (Cease), or the quaternary-ammonium product KleenGrow.

### Diagnostic Procedures

To avoid confusing symptoms of abiotic injury or fungal infection to those of bacterial leaf spot, the shape of the lesion is important to note. The lesion shape of bacterial leaf spot is defined by leaf veins and veinlets during early stages of development. Thus, lesions are angular rather than having the round shape typical of fungal leaf spots or injury caused by some pesticides or other chemical sprays. In some hosts, angular leaf spots can be caused by other factors such as downy mildews and foliar nematodes.

A simple slide making protocol can be followed to observe bacterial streaming from infected leaf tissue under a bright- or dark-field microscope (~100x). This can also be done with vascular tissue. Streak-plate isolations can be made from expanding lesions or vascular tissue on sucrose peptone agar (SPA) and yeast dextrose calcium carbonate agar (YDC). Allow the bacteria to grow for at least 48 hours at room temperature and examine periodically for colony growth before sub-culturing. Colony color and appearance, Gram reaction (using a 3% KOH solution) and the oxidase reaction are easy tests for preliminary confirmation when a Xanthomonas species is suspected.

When a mix of bacteria is recovered, a bioassay for the hypersensitive response can be used to determine which isolates warrant further testing. Aqueous suspensions with large numbers of bacterial cells are infiltrated into tobacco or pepper leaves (detached or in planta). Xanthomonas species do not provoke an HR as quickly as pathogenic pseudomonads, so the test should be observed for at least 48 hours. Note that if the bacterium happens to be compatible with the inoculated host, disease symptoms, rather than an HR, will develop.

Further identification can be conducted using fatty acid methyl ester (FAME) analysis and carbon source utilization tests such as the Biolog Identification system. It is important to note that in the

Biolog Gen III database (version 2_6_1_08.15G) all species are listed as Xanthomonas campestris, and only the following 13 pathovars are included: begoniae A, begoniae B, campestris, carotae, dieffenbachiae, juglandis, malvacearum, pelargonii, phaseoli, poinsettiicola, raphani, translucens, and vesicatoria.

ELISA tests are available from Agdia and AC Diagnostics for pathovars pelargonii, begoniae, and dieffenbachiae, and from AC Diagnostics for pv. hyacinthae. The test for pv. pelargonii is available in a lateral flow format from Agdia. Agdia also offers a generic ELISA test for Xanthomonas. ELISA testing can be performed on bacterial cultures or directly on plant tissue. Single gene sequencing (e.g. partial sequencing of the Gyrase B and other genes), once fully validated, and is likely to offer rapid and accurate identification of Xanthomonas species and pathovars.

## Erwinia Amylovora

Erwinia amylovora is the type species of the genus Erwinia that belongs to the family Enterobacteriaceae. As is common in bacterial taxonomy in recent years, there have been a continuous description of novel Erwinia species as well as some species have been reclassified and transferred to other genera. Most members of this genus cause diseases in plants and, historically, it is important to remember that E. amylovora was the first bacterium demonstrated to cause disease in plant, a discovery made in the late 1800s, at the same period as a similar discovery with human and animal diseases.

E. amylovora causes fire blight, a devastating plant disease affecting a wide range of host species within the Rosaceae subfamily Spiraeoideae, and is a major global threat to commercial apple and pear production. Moreover strains infecting plants in the genus Rubus belonging to the subfamily.

Among the limited number of control options currently available, prophylactic application of antibiotics (e.g., streptomycin or oxytetracycline) during the bloom period appears most effective. However, regulatory restriction, public health concerns, and pathogen resistance development severely limit the long-term prospects of antibiotic use. Biological control measures may offer promising alternatives to minimize or even substitute the use of antibiotics, and to mitigate occurrence of resistance.

In the pathogenesis of E. amylovora, pathogen cells enter plants through the nectarthodes of flowers and other natural openings, such as wounds, and are capable of rapid movement within plants and the establishment of systemic infections. In susceptible hosts, bacteria first move through the intercellular spaces of parenchyma and, in a later stage, in the xylem vessels, thus provoking extensive lesions, and sometimes complete dieback of the tree, under favorable climatic conditions. The diseased parts of the plant become brown or black, as if they had been swept by fire.

Many virulence determinants of E. amylovora have been characterized, including the Type III secretion system (T3SS), the exopolysaccharide (EPS) amylovoran, biofilm formation, and motility. To successfully establish an infection, E. amylovora uses a complex regulatory network to sense the relevant environmental signals and coordinate the expression of early and late stage virulence factors involving two component signal transduction systems, bis-(3′-5′)-cyclic di-GMP (c-di-GMP) and quorum sensing.

## Virulence Factors

E. amylovora is highly virulent and capable of rapid systemic movement within plant hosts and of rapid dissemination among rosaceous species, including apple and pear trees, when environmental conditions are favorable. The internal movement of the pathogen through the vascular system of plants and the ability of the pathogen to infect flowers, actively growing shoots, and rootstocks makes the management of fire blight difficult. It has been shown that two major virulence determinants are required for E. amylovora to infect and cause disease on host plants: the EPS amylovoran and the Hrp type III secretion system (T3SS).

## Exopolysaccharides (EPS) Amylovoran and Levan

Exopolysaccharides (EPS) have been suggested to play a key role in bypassing the plant defense system, in disturbing and obstructing the vascular system of the plant and in protecting the bacteria against water and nutrient loss during dry conditions. EPS is an important element in the biofilm formation of E. amylovora, enabling the bacteria to attach to several surfaces and to each other. Amylovoran has been shown to be the main factor necessary for biofilm formation, and levan is a contributing factor, the quantity of amylovoran produced by individual E. amylovora strains being correlated with the degree of virulence.

Amylovoran is a polymer of a pentasaccharide repeating unit that generally consists of four galactose residues and one glucuronic acid residue. The molecular size of amylovoran is influenced by several environmental conditions and cell-metabolism-related factors. The strains of E. amylovora that do not have the capacity to produce amylovoran are non-pathogenic and are unable to spread in plant vessels.

The amylovoran synthesis (ams) gene cluster involved in the biosynthesis of amylovoran produces 12 ams-encoded gene products (AmsA to AmsL). AmsC, AmsH, and AmsL are believed to be involved in oligosaccharide transport and assembly, while AmsA possesses a tyorisine kinase activity. AmsB, AmsD, AmsE, AmsG, AmsJ, and AmsK proteins appear to play a part in annealing the different galactose, glucuronic acid, and pyruvyl subunits to the lipid carrier in order to form an amylovoran unit. AmsF may process newly synthesized repeating units and/or be involved in their polymerization by adding them to an existing amylovoran chain. Finally, AmsI seems to have a distinct function in recycling of the diphosphorylated lipid carrier after release of the synthesized repeating unit.

Levan is another EPS produced by E. amylovora, considered as a virulence factor. It has been shown that the lack of levan synthesis can result in a slow development of symptoms in the host plant. The specific role of levan in pathogenesis is still unknown and deserves further research. Interestingly, in recent studies, it has been demonstrated that the depolymerase (DpoL1) encoded by the T7-like E. amylovora phage L1 efficiently degrades amylovoran. Exposure of the bacteria to either L1 phage or recombinant DpoL1 led to EPS degradation and a phenotype similar to the one observed with EPS mutants. The enzyme strips the cells, thereby paving the way for infection by providing access of L1 phage to the cell wall, and also facilitating the infection by other phages, as Dpo-negative Y2 phages.

## Type III Secretion System (T3SS)

The type III secretion system (T3SS) is one of the important virulence factors used by E. amylovora in order to successfully infect its hosts. As with other Gram negative phytopathogenic bacteria, E. amylovora uses this evolutionarily conserved secretion system to export and deliver effector proteins into the cytosol of host plant cells through a pilus-like structure, which forms the central core element of T3SS.

E. amylovora pathogenicity relies on a type III secretion system and on a single effector DspA/E. This effector belongs to the widespread AvrE family of effectors whose biological function is not fully understood. T3SS is composed of a large, cylindrically shaped macromolecular complex organized into a series of ring-like structures with inner rings, outer rings and neck structure. It is embedded in the inner and outer membrane of the bacteria, while spanning the periplasmic membrane and extending into the extracellular environment with a pilus filament. It has been shown that T3SS occurs only at the site of Hrp pilus assembly and that pilus guides the transfer of effector proteins outside the bacterial cell, favoring the "conduit/guiding filament" model. The T3SS of plant-pathogenic bacteria is mainly made out of Hrc proteins, encoded by hrp-conserved (hrc) genes among plant-pathogenic bacteria and Hrp proteins, encoded by hypersensitive response and pathogenicity (hrp) genes. In E. amylovora, hrc and hrp genes are clustered in a pathogenicity island, which contains four regions, i.e., an hrc/hrp region, an Hrc effectors and elicitors region, an Hrp-associated enzymes region, and an island transfer region.

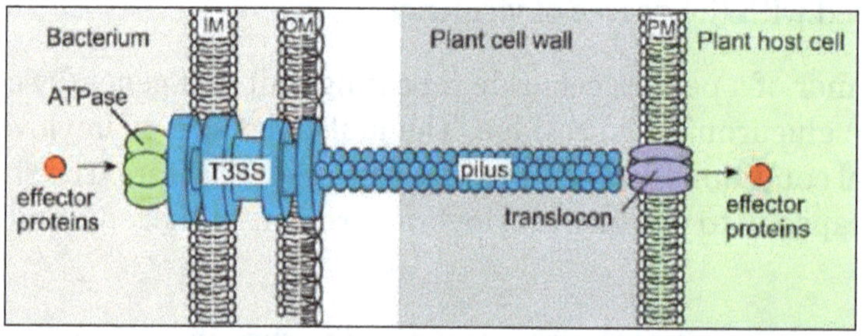


Schematic representation of the T3SS from plant pathogenic bacteria.

## Curli

Curli are the major proteinaceous component of a complex extra-cellular matrix produced by many Enterobacteriaceae, including E. amylovora. Curli fibres are known to be involved in adhesion to surfaces, cell aggregation and biofilm formation. Curli can also mediate host cell adhesion and invasion and they stimulate the host inflammatory responses. The structure and biogenesis of curli are unique among the bacterial fibres that have been described to date, belonging to a growin gclass of fibres, known as amyloids, which are associated with diverse neurological diseases in humans. Curli fibres are 4–6 nm-wide and, like other amyloid fibres, are β sheet-rich self-assembling protein polymers that are resistant to chemical and temperature denaturation, and to digestion by proteinases. Recent results have showed reductions in virulence due to deletion of a regulator of curli genes, crl, suggesting that not only functional attachment but also mature biofilm formation is needed for full virulence in the host. Further research is necessary to charaterize the chemical structures and genes encoding this structure and its specific role in pathogenicity, as it is also in process for other Enterobacteria as E. coli or Salmonella.

## Biofilm Formation

In addition to the specific role of the exopolysaccharides amylovoran and levanin in the biofilm formation of E. amylovora, a recent study has suggested that type I fimbriae, flagella, type IV pili, and curli of E. amylovora may contribute to biofilm formation in static and flowing environments and that defects in many of the genes encoding these appendages result in decreased virulence in planta.

E. amylovora forms a biofilm in vitro and in planta. Pathogenesis and biofilm formation appear to be linked, but without identifying genes encoding traits independent of amylovoran production, the mechanistic role in virulence of biofilm formation in E. amylovora could not be studied. Interestingly, mutants with reduced biofilm-formation ability appear unable to successfully establish large populations in apple xylem. Colonization of xylem is critical to the systemic movement of the pathogen through plants. Therefore, biofilm-deficient mutants remain localized within an inoculated leaf and are strongly impaired in the ability to invade the rest of the plant.

In a study using a bioinformatic approach and the recently sequenced genome of E. amylovora, genes encoding putative cell surface attachment structures were identified. A time course assay indicated that type I fimbriae function earlier in attachment, while type IV pilus structures appeared to function later in attachment. Deficiencies in type I fimbriae lead to an overall reduction in E. amylovora virulence. By deletion of individual genes and gene clusters and using a combination of in vitro attachment assays and plant virulence assays it was demonstrated that multiple attachment structures are present in E. amylovora and play a role in mature biofilm formation, which is critical to pathogenesis and systemic movement in the host. In contrast, a fully functional biofilm was not necessary to survival and growth in planta.

Although the mechanistic details behind biofilm formation remain largely unknown and genetic and structural analyses should be done for full characterization of these structures, it is suggested that they are formed in response to environmental triggers and quorum sensing signals.

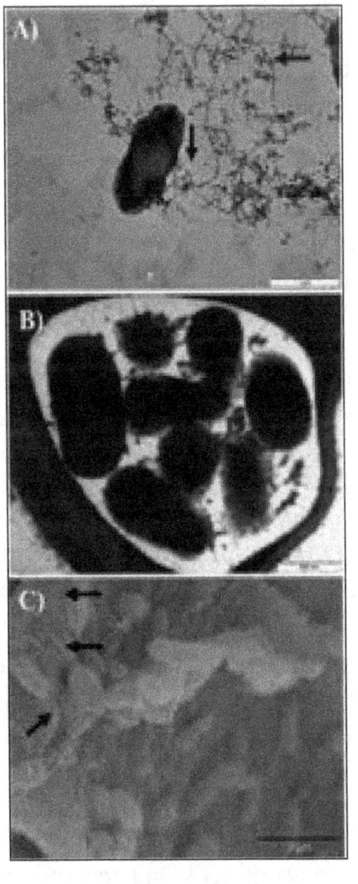


Figure above shows putative attachment structures of E. amylovora. (A) TEM imaging of a planktonic E. amylovora cell grown in broth culture and negatively stained. Peritrichous flagella are indicated by arrows (scale: 1 μm); (B) TEM image of E. amylovora in planta. Putative attachment structures connect bacterial cells to host cells (scale: 1 μm); (C) SEM image of E. amylovora cells found within a biofilm, with multiple appendages that protrude from the bacterial cell and attach to the host surface, as indicated by the arrows (scale: 2 μm).

## Motility

E. amylovora is a nonobligate pathogen able to survive outside the host under starvation conditions, allowing its spread by various means, such as rainwater. In fact, motility mechanisms have been shown to be of relevance for the survival of the bacteria ouside the host and also for the attachment to the host cell surfaces, together with other virulence factors as type IV pili, type I fimbriae and curli by biofilm formation.

In a study where two of the four gene clusters encoding the production and regulation of flagella in E. amylovora were deleted, various effects of flagella on biofilm formation and virulence were demonstrated. Results obtained also showed that flagellum production appeared to be controlled by multiple gene clusters that may function independently. Consequently, it was suggested that the flagella of E. amylovora have multifaceted functions in the biofilm formation process.

Swarming motility has also been described in E. amylovora, although, as in other Enterobacteria, its role remains unknown, requiring further research E. amylovora responses to starvation revealed that cells lost motility but most of them exhibited long peritrichous flagella attached to the bacterial surface. Motility inactivation during starvation would be reversible since flagella could enable a rapid recovery of motility under favorable conditions.

## Lipopolysaccharide (LPS)

LPS is a factor that has been described recently to be involved in the virulence of E. amylovora, although the complete chemical structure of LPS and its relationship with the pathogenicity has not yet been described.

Interestingly, in Erwinia carotovora, a comprehensive physicochemical analysis of highly purified mono- and bis-phosphoryl hexa- to heptaachyl and pentaacyl lipid. The distinct differences observed between different Lipid A structures could explain differences in the biological activity, such as the cytokine inducing activity.

At the genetic level, comparative genomic analysis revealed differences in the LPS biosynthesis gene cluster between the Rubus-infecting strain ATCC BAA-2158 and the Spiraeoideae-infecting strain CFBP 1430 of E. amylovora. These differences were restricted to the core region that could be involved in a process of adaptation to the new host. Genetic differences observed in the LPS biosynthetic gene cluster corroborate rpoB-based phylogenetic clustering of E. amylovora into four different groups and enable the discrimination of Spiraeoideae- and Rubus-infecting strains.

## Metalloprotease PrtA

A metalloprotease with a molecular mass of 48 kDa secreted by E. amylovora has been characterized. The protease is apparently secreted into the external medium through the type I secretion pathway via PrtD, PrtE, and PrtF, which share more than 90% identity with the secretion apparatus for lipase of S. marcescens.

A gene cluster encodes four genes connected to protease expression, including a structural gene (prtA) and three genes (prtD, prtE, prtF) for secretion of the protease, which are transcribed in the same direction. The organization of the protease gene cluster in E. amylovora is different from that in other Gram-negative bacteria, such as Erwinia chrysanthemi, Pseudomonas aeruginosa, and Serratia marcescens. On the basis of the conservative motif of metalloproteases, PrtA has been identified to be a member of the metzincin subfamily of zinc-binding metalloproteases, and has been confirmed to be the 48 kDa protease on gels by sequencing of tryptic peptide fragments derived from the protein.

In a protease mutant created by mutations with Tn 5–insertions in the prtD gene, the lack of protease reduced colonization of an E. amylovora secretion mutant, labeled with the gene for the green fluorescent protein (gfp) in the parenchyma of apple leaves.

Further research is necessary, however, to assess its structural characteristics, activity, substrates, the specific role of this metalloprotease in the pathogenic mechanisms of E. amylovora, and also the possible practical applications of this protein, based on the differences observed with other Gram negative bacteria, as already proposed for this interesting wide family of proteins.

## Iron-Scavenging Siderophore Desferrioxamine

Under iron-limiting conditions, E. amylovora CFBP 1430 produces hydroxamate-type siderophores, characterized as cyclic desferrioxamines (DFOs), mostly DFO E and the specific receptor FoxR, a 70,000-Da protein needed for the passage of ferric complexes across the outer membrane.

Several iron uptake negative mutants of E. amylovora CFBP 1430 isolated by insertional mutagenesis have been identified. In mutant VD61, the mutation (dfo-61:: MudIIpR13) disrupts the DFO biosynthetic pathway; in VD17, the mutation (foxR-17::MudIIpR13) affects the synthesis of the FoxR receptor and the mutant accumulates DFO in the external medium because of its failure to transport back the DFO ferric complex. In the pathogenic analysis of these mutants, results obtained demonstrated that, in E. amylovora, production of DFOs is a critical function for bacterial iron acquisition during pathogenesis, as well as for bacterial induction of electrolyte leakage from plant cells. Based on this, a dual role of DFO during plant/E. amylovora interactions has been proposed, which deserve further research.

## Multidrug Efflux Pump AcrAB

During pathogenesis, E. amylovora is exposed to a variety of plant-borne antimicrobial compounds, which in the case of plants of Rosaceae, many are constitutively synthesized as isoflavonoids. Then, bacterial multidrug efflux transporters, which mediate resistance toward structurally-unrelated compounds, might confer tolerance to these antimicrobial compounds (phytoalexins). In this regard, it has been observed that clonation of the acrAB locus from E. amylovora, encoding a resistance nodulation division-type transport system, in E. coli conferred resistance to hydrophobic and amphiphilic toxins. An acrB-deficient E. amylovora mutant was impaired in virulence on apple rootstock MM 106 and was susceptible toward extracts of leaves of MM 106, as well as to the apple phytoalexins phloretin, naringenin, quercetin, and (+)-catechin. These results strongly suggest that the AcrAB transport system plays an important role as a protein complex required for the virulence of E. amylovora in resistance toward apple phytoalexins and that it is required for successful colonization of a host plant.

## Other Virulence Factors

Other virulence factors of E. amylovora have been proposed, as the case of the sorbitol transporter, encoded by the srl operon. It has been shown that expression of the srl operon in E. amylovora is high in the presence of sorbitol in medium and is repressed by glucose. Mutants with a sorbitol deficiency were still virulent on slices of immature pears, but were unable to cause significant fire blight symptoms on apple shoots. Since sorbitol is used for carbohydrate transport in host plants of E. amylovora, this sugar alcohol has been proposed to be an important factor in determining host specificity for the fire blight pathogen.

Moreover, rosaceous plants also contain sucrose as storage and transport carbohydrates. The regulation and biochemistry of sucrose metabolism of E. amylovora has also been studied, demonstrating that sucrose mutants (in the scr regulon), created by site-directed mutagenesis, did not produce significant fire blight symptoms on apple seedlings, thus, indicating the importance of sucrose metabolism for colonization of host plants by E. amylovora.

## Genetics

The genome of E. amylovora consists of a circular chromosome of 3,805,874 bp and two plasmids: AMYP1 (28,243 bp), also reported as PEA29, and the larger plasmid AMYP2 (71,487 bp), also named pEA72. The small size of the E. amylovora genome (3.8 Mb) in comparison with most free-living enterobacteria, including plant pathogens, with genomes of 4.5–5.5 Mb, and the preponderance of pseudogenes, genome reduction can have occurred via mutational inactivation and subsequent deletion and, as a consequence, has led to the loss of most of the genes involved in anaerobic respiration and fermentation found in typical, related enterobacteria, with the consecuent reduction of the capacity to live in anaerobic environments. The genome sequence of E. amylovora has revealed clear signs of pathoadaptation to the rosaceous plant environment. For example, T3SS-related proteins are more similar to proteins of other plant pathogens than to proteins of closely related enterobacteria.

Comparative genomic analysis of 12 strains representing distinct populations (e.g., geographic, temporal, host origin) of E. amylovora has allowed describing the pan-genome of this species. The pan-genome contains 5751 coding sequences and is a highly conserved relative to other phytopathogenic bacteria, comprising on average 89% conserved core genes. While the chromosones of Spiraeoideae-infecting strains are highly homogeneous, greater genetic diversity was observed between Spiraeoideae- and Rubus-infecting strains (and among individual Rubus-infecting strains).

Using molecular approaches based on the study of repetitive elements, such as Multiple Loci Variable Number of Tandem Repeats Analysis (MLVA) or sequencing of Clustered Regularly Interspaced Short Palindromic Repeats (CRISPR), this diversity has also been especially observed among strains isolated from Rubus plants and, to a lesser extent, in Spiraeoideae-infecting strains.

In comparison with other plant pathogens, however, E. amylovora has relatively low genetic diversity, with a limited genetic recombination, this being associated with its exposure to limited selection pressures due to pome fruit breeding strategically favoring only high-valued varieties, which are often highly susceptible to fire blight.

On the basis of the current available data, it has been hypothesized that the critical event for adaptation to Rubus spp. took place after species separation of E. amylovora and E. pyrifoliae, as the Spiraeoideae infecting isolates of E. amylovora and E. pyrifoliae (including Japanese strains), as well as E. tasmaniensis and E. piriflorinigrans, all sharing the Spiraeoideae-type LPS biosynthetic cluster. Interestingly, genome sequencing of three Mexican E. amylovora strains have revealed an rpsL chromosal mutation conferring high-level of streptomycin resistance. This chromosomal mutation is the predominant E. amylovora mechanism, typically appearing after years of intensive application with long persistence in populations, explaining the observed inefficacy of streptomycin in Mexico and underscoring the need of revision of pesticide use strategies to avoid similar resistance evolution against other antimicrobial drugs as oxytetracycline or gentamicin.

The expression of genes related to starvation, oxidative stress, motility, pathogenicity, and virulence have been detected during the entire experimental period with different regulation patterns observed during the first 24 h. Further, starved cells remained as virulent as nonstressed cells. Overall, these results provide new knowledge on the biology of E. amylovora under conditions prevailing in nature, which could contribute to a better understanding of the life cycle of this pathogen.

No major differences among 12 strains of E. amylovora were reported in the amylovoran biosynthesis cluster (>98% amino acid indentity across the whole region). Variation, however, has been observed in the Hrp cluster, a pathogenicity island that encodes the hypersensitive response and pathogenicity (hrp) T3SS and the majority of known T3SS effector proteins. Variation was identified in HrpK, the putative chaperones OrfA and OrfC (which varied between host specific grouping of Rubus- and Spiraeoideae-infecting strains) and, more significantly, Eop 1, which has been shown to function as a host-limiting factor.

Three gene clusters related to the Type 6 Secretion System (T6SS) have also been identified in E. amylovora, although their exact role in this species is unknown and should deserve further research. In other bacteria, T6SS has been shown to play a significant role in bacterial-bacterial and bacterial-host interactions, suggesting a role for niche specialization.

## Regulation of Virulence Factors

### Two Component Transduction Systems

The phoPQ two component transduction systems has been genetically characterized in E. amylovora and it has been demonstrated that this system in E. amylovora plays major roles in virulence on immature pear fruit and in regulating amylovoran biosynthesis and swarming motility. It has also been shown that phoPQ mutants were more resistant to strong acidic conditions (pH 4.5 or 5) than that of the wild-type (WT) strain, suggesting that this system in E. amylovora may negatively regulate acid resistance gene expression. Furthermore, the PhoPQ system negatively regulated gene expression of two novel T3SS in E. amylovora. These results are in contrast to those reported for the PhoPQ system in Salmonella and Xanthomonas, where it positively regulates T3SS and acid resistance. In addition, survival of phoPQ mutants was about 10-fold lower than that of WT when treated with cecropin A at pH 5.5, suggesting that the PhoPQ system renders the pathogen more resistant to cecropin A.

These results suggest that the the PhoPQ system may act as a part of a regulatory network that governs E. amylovora in a wide variety of cellular functions, thus enhancing its survival under the changing environments of different hosts or in different environments, such as the acidic plant apoplast, a major barrier for plant pathogenic bacteria to cause disease.

## Small RNA

Recently, different regulatory small RNAs (sRNAs) requiring the RNA chaperone Hfq for both stability and functional activation have been identified in E. amylovora, and further, being some of them involved in the regulation of various virulence traits including motility, amylovoran EPS production, biofilm formation, and T3SS. This discovery is of relevance since, although sRNAs have been increasingly recognized as pivotal regulators in bacteria, genome-wide identification of sRNAs has only been performed in a limited number of bacteria.

## Quorum Sensing

Quorum sensing systems have been described in E. amylovora, both QS-1, relying on an N-homoserine lactone autoinducer-1 signal (AI-1) and QS-2 reliant on LuxS as the enzyme responsible for signal (AI-2) production. However, a global regulatory function has not yet been explored. LuxS did not affect quorum sensing in E. amylovora, recent results have indicated that some strains of E. amylovora have luxS-dependent AI-2 activity, and that inactivation of luxS affects some phenotypes, including virulence in planta, although the molecular basis of luxS-dependent regulation remains to be determined. It was observed that AI-2 produced in E. amylovora activated the expression of bioluminescence in the Vibrio harveyi BB170 reporter strain. Findings showed that AI-2 in Ea1665 was dependent on EaluxS because expression of EaluxS in the luxS frame-shift mutation of E. coli DH5α demonstrated its importance in AI-2 synthesis. The pattern of AI-2 production with respect to growth phase and environmental changes deserves further research, and may give some insight into the function of this signaling system.

Moreover, using comparative genomic analyses to search homologs of genes involved, several genes were identified corresponding to the Autoinducer-3/Epinephrine/Norepinephrine signaling system in E. coli. Particularly, the putative homologs to the two component system QseEF and the transcriptional regulator QseA are highly conserved in E. amylovora. Gene-knockout experiments revealed that the QseEFEa system is involved in the regulation of swarming motility, biosynthesis of amylovoran, and biofilm formation.

## c-di-GMP

It has been shown that E. amylovora encodes eight genes involved in the biosynthesis and/or degradation of cyclic-di-GMP (c-di-GMP), a second messenger signaling compound. Individual knockout mutations in each of these genes revealed that c-di-GMP positively regulated biofilm formation and negatively regulated motility. These results demonstrated that E. amylovora possesses several regulatory networks that govern the expression of virulence genes.

## ppGpp

In a recent study, the role of the bacterial alarmone ppGpp in activating the T3SS and virulence of

E. amylovora was investigated using ppGpp mutants generated by Red recombinase cloning. The virulence of a ppGpp-deficient mutant (ppGpp(0)), as well as a dksA mutant of E. amylovora, was completely impaired, and bacterial growth was significantly reduced, suggesting that ppGpp is required for full virulence of E. amylovora. Expression of T3SS genes was greatly downregulated in the ppGpp(0) and dksA mutants. Western blotting showed that accumulations of the HrpA protein in the ppGpp(0) and dksA mutants were about 10% and 4%, respectively, of that in the wild-type strain. Furthermore, higher levels of ppGpp resulted in a reduced cell size of E. amylovora. Moreover, serine hydroxamate and α-methylglucoside, which induce amino acid and carbon starvation, respectively, activated hrpA and hrpL promoter activities in hrp-inducing minimal medium. These results demonstrated that ppGpp and DksA play central roles in E. amylovora virulence and indicated that E. amylovora utilizes ppGpp as an internal messenger to sense environmental/nutritional stimuli for regulation of the T3SS and virulence.

## Type 3 Effectors

There are several effectors in E. amylovora that have been described, although recent findings have been focused on DspA/E since it has been shown that the ability of E. amylovora to promote disease mainly depends on this single injected T3E. DspA/E belongs to the widespread AvrE family of type III effectors that suppress plant defense responses and promote bacterial growth following infection. In recent studies, expression of dspA/E in the yeast Saccharomyces cerevisiae inhibited cell growth, this effect associated with perturbations of the actin cytoskeleton and endocytosis, activation of phosphatase 2A, and downregulation of the sphingolipid biosynthetic pathway leading to growth arrest.

In non-host Arabidopsis thaliana leaves, DspA/E was required for transient bacterial growth, as an E. amylovora dspA/E mutant was unable to grow. Study of transgenic Arabidopsis lines expressing DspA/E indicated that DspA/E promotes modifications of plant cell metabolism among which the repression of protein synthesis could be determinant in the facilitation of necrosis and bacterial growth.

## Hrp X/Y, HrpS and HrpL Cascade Leading to Activation of HrpT3SS

Two regulatory components, hrpX and hrpY, of the hrp system of Erwinia amylovora have been identified. The hrpXY operon is expressed in rich media, but its transcription is increased threefold by low pH, nutrient, and temperature levels, conditions that mimic the plant apoplast. hrpXY is autoregulated and directs the expression of hrpL, which, in turn, activates transcription of other loci in the hrp gene cluster.

The deduced amino-acid sequences of hrpX and hrpY are similar to bacterial two-component regulators including VsrA/VsrD of Pseudomonas (Ralstonia) solanacearum, DegS/DegU of Bacillus subtilis, and UhpB/UhpA and NarX/NarP, NarL of Escherichia coli. The N-terminal signal-input domain of HrpX contains PAS domain repeats.

hrpS, located downstream of hrpXY, encodes a protein with homology to WtsA (HrpS) of Erwinia (Pantoea) stewartii, HrpR and HrpS of Pseudomonas syringae, and other $\sigma^{54}$-dependent, enhancer binding proteins. Transcription of hrpS also is induced under conditions that mimic the plant apoplast. However, hrpS is not autoregulated, and its expression is not affected by hrpXY. As a member of the NtrC family, HrpS is unusual in that it lacks a long N-terminal receiver domain.

The structure of the input domain of E. amylovora HrpX appears to be exceptional, compared with sensor proteins involved in other type III systems, which contain two transmembrane regions and a periplasmic domain. Thus, at least two types of transmitter-receiver systems appear to have evolved for control of type III systems in response to environmental stimuli in hosts.

## Regulatory Cascade for Amylovoran Synthesis

The biosynthesis of amylovoran is regulated by another two-component signal transduction system, the RcsCDB phosphorelay system, which has been found to be essential for virulence in E. amylovora.

## Survival of Plant Pathogenic Bacteria

### Short-Term Survival

Plant pathologists are mostly concerned with the practical effects of long-term survival. However, it is appropriate that attention be given here to the survival of metabolically and physically active bacteria for shorter periods-hours to days. Physical, chemical, and microbiological factors affecting short-term survival also affect long-term survival, but information about long-term survival is much sketchier. A brief discussion of inanimate and animate factors affecting short-term survival of bacteria is appropriate.

## Inanimate Factors Affecting Survival

Water governs the life of bacteria in vital ways and has a marked effect on short-term survival. Free water probably is necessary for the multiplication of bacteria in general, and it is of course required for swimming. Metabolically active forms of most bacteria are killed by drying, but there are differences in degree owing to species and conditions of drying.

Active plant pathogenic bacteria existing independently probably die quickly when they are dried, as would be expected of non-sporing bacteria in general. There appears to have been no systematic study of plant pathogenic bacteria in this respect, however. Some '70 years ago, Jones and Smith placed active cells of pathogens in drops on cover slips. Slips were dried and assayed later in vitro. Results indicated that season-to-season survival was unlikely and that survival depended on conditions of drying. Jones found that a potato soft rot bacterium he studied died within minutes when water suspensions were dried. Work by Kauffman and Leben with Pseudomonas glycinea suspended in buffer demonstrated that ca. l on cells per ml. of this pathogen died virtually as soon as drops of suspension dried on cover slips. Another recent example of sensitivity to desiccation is that of Kikumoto and Sakamoto, who found that Erwinia aroideae cells died rapidly when soil particles colonized by this pathogen were dried.

The relative humidity during drying influences survival of many kinds of bacteria. For example, Erwinia amylovora, in common with the intensively studied Escherichia coli, survived best at 40-90 % relative humidity in air-borne particles. A number of works indicated that a low relative humidity limits multiplication of many kinds of macro- and micro-flora on the plant surface, including plant pathogenic bacteria. On the other hand, high humidity favors epiphytic growth in general.

Swimming may be far more important for survival of plant pathogenic bacteria than present information suggests. Movement toward or away from a stimulus is an important fitness factor for many organisms which swim. Many pathogenic bacteria are motile, and some experiments have indicated they move toward seemingly favored positions. It is not difficult to envision a bacterium and its progeny swimming in a drop of water on a leaf toward a source of nutrients (attractants) and that this location may provide a survival site. Repellants as well as attractants may help "guide" swimming bacteria. The swimming of a virulent Pseudomonas solanacerarum isolates appeared to be a fitness factor not for in vivo growth but for in vitro growth, because Kelman and Hruschka found that virulent isolates were not motile. The subject of swimming of plant pathogens needs much more investigation. Numbers of plant pathogenic bacteria exposed on the above-ground parts of plants probably are quickly reduced by the ultraviolet (UV) rays in sunlight. Inactivation by UV is influenced by relative humidity and other factors. Death from UV must be very large when pathogens washed from aerial lesions are spread widely during a wind and rain storm. Many years ago, Jones discovered the sensitivity of the soft rot bacterium he studied to sunlight and recommended exposing potato tubers to sunlight as well as to drying conditions for disease control.

Little work has been done with plant pathogens, however. In addition to water and sunlight, two other inanimate factors, temperature and the chemical environment at the survival site, undoubtedly have a profound influence on pathogen survival. They also influence other microorganisms, especially if water is present and the chemicals serve as nutrients for these organisms. The interactions among inanimate factors are varied, complex, and little studied with respect to their influence on the survival of pathogens.

## Animate Factors Affecting Survival

Prior to and during infection, and with the aging of a lesion, pathogenic bacteria are subject to manifold pressures exerted by other microorganisms. Probably the only time when a pathogen is free, or nearly free, from influences of these other organisms is during the early invasive growth period within plant tissue. As a lesion ages, especially in a moist environment and especially in contact with soil, omnipresent nonpathogenic microorganisms begin the decomposition of the lesion and the masses of pathogen cells it contains. In addition to the micro-flora, many kinds of fauna may be present, and some of these consume bacterial pathogens. It is probable that in regions where plant tissues decompose rapidly, the survival time for a pathogen is shorter than in regions where decomposition of organic materials takes place more slowly.

The life expectancy of metabolically active, independent cells of pathogenic bacteria likely is a short one. Consider the soil environment. Assume that a bacterium began its independent existence by being washed from a leaf lesion into the soil or by swimming from a colony near a root. Microorganisms in the soil are largely inactive, owing to the sparsity of energy sources and to other unfavorable factors. More recently, Brown presented additional evidence that bacterial members of the soil flora are in a state of reduced metabolism most of the time and refers to this condition as "bacteriostasis", a term parallel to "fungistasis".

An immigrating pathogen would encounter these adversities. If an energy source should become available, very likely other organisms would be better able to use it; because many plant pathogens are not nutritionally versatile. Other organisms would possess a greater "relative competitive advantage" with respect to energy sources, to use the term of Cook and Papendick. Thus, it seems likely that unless a suitable niche could be found quickly, immigrating cells of bacterial pathogens in the soil environment would expire within a short time.

Yet some pathogens are "soil borne" and do persist m soil for years. These are mentioned below. Life expectancy of active, independent cells of a pathogen m an aerial environment also would be expected to be short. In Ohio climate, bacteria washed from an active lesion on a leaf to another area of the foliage probably would die soon, owing to desiccation or UV irradiation. If the cell did not meet this end, it then would face the competition of other organisms already living on the plant. Crosse, Gibbins, Last and Warren, and Leben has written about the relationships among epiphytic organisms. A notable symposium covering many aspects of the microbial ecology of the leaf surface was held in 1970 in Newcastle.

## Constructs

Three constructs are offered as a framework for thinking about the survival of plant pathogenic bacteria:

- Long-term survival of pathogens in nature takes place only in association with living or dead plant tissues. This is a "critical trait" which allows pathogens to survive in the face of recurrent or occasional stresses.
- Long-term survival is not likely to take place unless cells of the pathogen are in aggregates or unless they are associated with living plant tissues in "protected positions".
- Pathogens in a state of reduced metabolism are more likely to survive than are active cells.

## Hypobiosis

Bacterial cells in a state of reduced metabolism are designated hypobiotic cells. Yet the concept bears special significance. In the hypobiotic state, microorganisms may live long periods without added nutrients and are more likely to survive the physical and chemical stresses causing death than when metabolic activity is high. Hypobiosis can be induced experimentally by low temperature, loss of water, increasing salt concentration, and a variety of other means.

It seems reasonable to conclude that pathogens surviving for long periods are in a hypo biotic condition, having arrived at that state as a result of the natural processes takjng place with the aging of diseased tissues. Hypobiotic cells would be the survivors in dry leaf, stem, and root lesions of annual plants: they would represent a small portion of the masses of cells once alive within the lesion. Hypobiotic cells are, of course, quite different from actively metabolizing ones, and they deserve much more study than has been given them. For example, cells of Pseudomonas aeruginosa were more sensitive to desiccation in the exponential growth phase than when cells were 7 days old. Normand et al. found that the morphology of Pseudomonas phaseolicola cells was different in the lesion center than at the lesion edges, where the bacteria were younger.

## Long-Term Survival

Some bacterial pathogens can survive for many ·years if they are in diseased plant tissues which are dry, as is well known. The surviving cells in these tissues, in addition to being in a hypobiotic state, probably are protected in varying degrees by the surrounding masses of dead bacterial and plant cells and by products of the pathogen-host interaction.

The filamentous form of exudate oozing from a cotyledon lesion incited by Pseudomonas glycinea. Exudates consist of masses of bacteria in a matrix.

On the other hand, some pathogens die when the lesion dries. If a pathogen survives in the dry lesion and the lesion is moistened and exposed to microbial decomposition, as in soil, it is anticipated that surviving cells surrounded by a quantity of diseased tissue would not be as readily decomposed as when imbedded in smaller amounts. Furthermore, cells within tissues would be less subject to decomposition than those in soft tissues.

Debris from diseased plants must always be considered a possible source for seasonal carryover. However, two questions should be asked. How likely is it that debris will decompose between seasons? If it is not decomposed, what is the likelihood that surviving pathogen cells will come in contact with susceptible plant tissue? In working with Pseudomonas glycinea, for example, it was found that overwintering in quantities of diseased leaves was poor in soil, better on sod, and better yet if diseased leaves were suspended in air. The differences are attributed to the amounts of moisture and the relative numbers of decomposing microorganisms at the three sites. With ordinary crop cultivation methods, survivors in diseased leaves probably would occur near susceptible seedlings rarely (if ever), particularly if crops were rotated.

On the other hand, diseased debris associated with seed likely would he kept dry and thus not subject to deterioration. In this condition, cells of a pathogen may well survive as long as the seed is viable. This is an important subject, because survival with seed is responsible for much damage produced by bacterial pathogens of annual crops.

Lesion exudate is another type of debris resulting from disease. Exudates consist of masses of pathogen cells in a matrix, which may harden. Thus with Erwinia amylovora, exudate in the form of strands may disseminate the pathogen via air currents within the orchard. Survival in dry strands for months has been reported. Exudates are produced in lesions incited by many pathogens. For example, when head-shaped exudates of Pseudomonas p,-Zycinea were moistened, they were dispersed and formed flakes when dried. Flakes could be disseminated locally by the wind. It is suspected that long-term survival of pathogens in exudates in nature is not common, though, because most exudates would be degraded by physical or chemical a-action, or they would become moistened and destroyed by micro-organism.

## Protected Positions

It is clear that many bacterial pathogens survive well ·under some conditions if they are in aggregates. As indicated, they probably die quickly as individuals in soil or on exposed sites above ground. However, recent evidence suggests that individuals or a few cells can survive for varying periods of time in association with the healthy living plant or plant part, at sites termed "protected". It is emphasized that plant tissues are healthy as far as can be determined.

The protected position hypothesis first was deduced by experiments in which water drops were placed on healthy cucumber leaves, and the areas covered by these drops were examined later for bacteria. Sizeable populations of nonpathogenic bacteria developed in 48 hours in many drop areas. In contrast, few bacteria were present in areas not covered by drops. It was concluded that the progenitors of these bacteria were in some sort of sheltered locations on the leaf surface before drops were placed over them. Probably they were hypobiotic. Blakeman and Blakeman and Fraser also obtained a rapid buildup of epiphytic bacteria in water drops on leaves of two plant species.

The bacteria just discussed were nonpathogenic. Recent work suggests that pathogenic bacteria also survive in protected positions on healthy leaves. Mew and Kennedy found that Pseudomonas glycinea multiplied and survived for at least 14 days on healthy soybean leaves in the greenhouse. Scherff and Kauffman and Leben confirmed these results. In the latter tests, three levels of relative humidity were used; multiplication and survival were observed at each level: low, medium, and high. Multiplication or survival at the low relative humidity was not expected because P. glycinea is sensitive to drying and epiphytic growth of all types is reduced by a low relative humidity. Consequently, it was concluded that the pathogen multiplied in a protected, moist site and survived at or near this location. Other workers have found pathogens surviving on healthy host or non-host leaves.

From these examples, one may extend the hypothesis of protected positions to other locations associated with the healthy living plant or plant part. For example, a few hypobiotic cells of a pathogen within a seed would be in a well-protected and significant survival site. Buds may be protected survival sites, and pathogens may survive within fruit tree stems. The short report of Meneley and Stanghellini indicated that the interior parts of a number of fresh vegetable products served as protected survival sites for bacteria which caused tissue decomposition when these products were warmed.

The rhizosphere also may serve as a protected position for the survival of some pathogenic bacteria. Pathogens have been found associated with apparently healthy roots of host and non-host plants. Possibly these pathogens are living in the mucigel surrounding the root. It seems significant that most pathogens do not survive well in soil, and the few that do persist, the soil-borne pathogens, possess wide host ranges.

Assume that a soil-borne pathogen is able to occupy a niche on the rhizoplane of healthy host or non-host plants. Pathogens at this location then could exist in soil for as long as there were roots of suitable species to provide sites for multiplication and survival. If suitable crop or weed plants were not grown, the pathogen would disappear. Unless plants became diseased, with the subsequent release of large amounts of inoculum, populations probably would remain at relatively low levels. A pathogen of this sort could be distributed in soil by contact between roots, by moving water and agricultural implements, by swimming, or by nematodes and other fauna.

A parallel with Rhizobia spp. is evident here. Rhizobia grow and survive in association with non-host as well as host roots, and it is suggested that they survive in the soil as long as these roots are present. If this suggestion is correct, an explanation is offered for the long-term disease potential of certain pathogens in the soil in the face of what is described as a hostile environment. Protected positions probably are most significant for low numbers of cells of either pathogenic or nonpathogenic bacteria. However, it is possible that large numbers also could survive at these locations.

## Contributions of the Pathogen Life Cycle Phases to Survival

The life cycle of a plant pathogenic bacterium may be divided into four phases or stages. These are the pathogenic, resident, saprophytic, and survival phases. The first three-pathogenic, resident, and saprophytic are growth phases. What is the contribution of each in producing surviving cells?

### Pathogenic Phase

As far as known, the pathogenic phase, in which there is a large increase in numbers of pathogen cells and the production of symptoms, contributes most of the cells entering a seasonal survival period. The larger the population of a pathogen entering the period, the greater the chances for survival, other conditions being equal. Large populations would be produced by compatible isolate-host combinations (those resulting in typical disease responses); incompatible combinations (resistance responses), which produce lower populations, would be much less likely to yield surviving cells.

### Resident Phase

A decade ago it was suggested, on the basis of work with Xanthomonas vesicatoria, that some pathogenic bacteria possessed a “resident phase” in their life cycle. This was defined as the capacity for multiplication on the surface parts of the healthy shoot system.

It now appears as a result of recent studies that a number of bacterial pathogens possess a resident phase, which may be associated with leaves, buds, or flowers of host or non-host plants. Earlier work is cited by Leben. Pseudomonas syringae which has a wide host range is the subject of most of these papers. Many leaf-spotting pathogens may be residents under some conditions. The increase of a pathogen in the absence of symptoms may be of epidemiological importance by serving to build up an inoculum as an immediate prelude to infection. That increase also could provide pathogen cells which survived unfavorable times. Probably survival sites would he at or near multiplication sites.

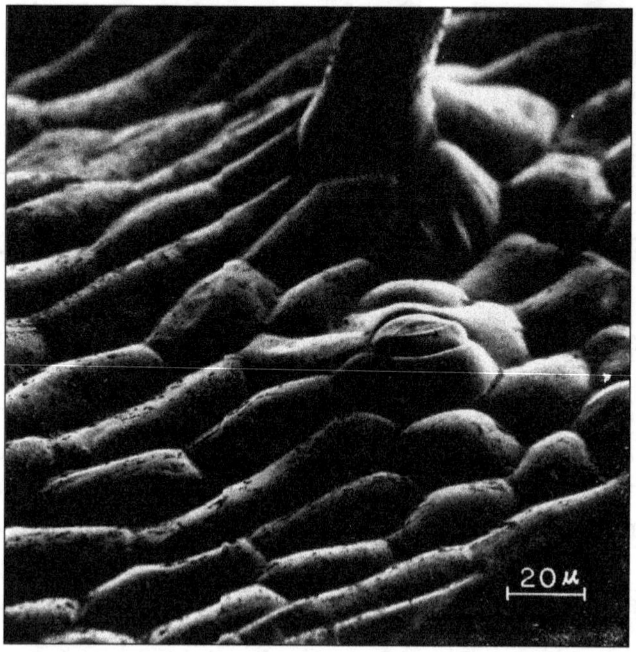


A scanning electron microscope view of the surface of a soybean stipule, showing pillow-shaped epidermal cells, the base of a trichome, and a stoma. Most of the small projections on the epidermal cells are aggregates of bacteria. These organisms also are found in abundance in depressions between epidermal cells, especially near the base of trichomes. These locations are possible protected survival positions.

The location of these sites is speculative, hut there are several possibilities. Leaves have been studied the most. In Ohio, relatively few non-pathogenic bacteria were detected on leaves of field plants. This suggests that these organisms and likely pathogenic bacteria as well, survive in protected positions. Probably there is a periodic (possibly diurnal) rise and fall in numbers of emigrants swimming or being washed from these sites, with population levels being governed primarily by the amount of water on the leaf, duration of periods of wetness, relative humidity, and exposure to sunlight. The leaf surface, especially within deep, moist depressions between epidermal cells, probably is a multiplication and survival site. The underside of the leaf would be expected to have more protected positions than the exposed upper surface.

The sub-stomatal chamber and associated tissues also are suspect locations. However, water does not readily enter stomata pores. Protected positions may be associated with trichomes or with hydathodes or other natural openings. These hypothetical locations could be occupied by pathogenic and nonpathogenic bacteria. In addition, pathogens may multiply and survive unnoticed as a result of an arrested infection of a few epidermal cells. More work is needed to understand what is taking place, particularly with pathogens. Buds also are worthy of comment as sites of multiplication and survival of pathogenic bacteria. In contrast to leaves in Ohio, the bud habitat (the gemmisphere) of some plants may carry a high population of a varied bacterial flora. The numbers and variety of non-pathogens in soybean buds, flowers, and young fruits were remarkable. Recently a Pseudomonas syringae type was found among the many bacteria in buds of healthy field soybean plants. Curiously, this pathogen produced progressive lesions on the soybean hypocotyl and cotyledon, but only incited a resistance response on leaves.

### Saprophytic Phase

Do plant pathogenic bacteria have a true saprophytic stage in nature? Are pathogens able to grow-multiply- on dead tissue derived from either host or non-host plants? Can pathogens multiply on other material in natural habitats? Some pathogens in aggregations readily survive in dead host tissue, as has been seen, but do they increase in these tissues? No direct evidence is known for the proposition that they do increase. On the other hand, there is much evidence that population decrease. With the soybean blight pathogen, for example, once leaf lesions became necrotic, the pathogen population decreased. This took place if leaves were left on the plant or if they were detached and placed in soil or other natural situations.

Probably saprophytic growth has been suspected to take place most often in soil. If the explanation proposed above for long-term persistence of soil-borne pathogens is accepted, saprophytism would not be essential for survival. In general, soil seems to be particularly inhospitable for bacterial pathogens, as has been mentioned. Consequently, it appears that true saprophytic growth in this medium is limited. If so, it would have little effect on survival. Obviously, more needs to be known about this difficult subject.

## Plant Pathogenic Virus

Viruses are infectious pathogens that are too small to be seen with a light microscope, but despite their small size they can cause chaos. The simplest viruses are composed of a small piece of nucleic acid surrounded by a protein coat. As is the case with other organisms, viruses carry genetic information in their nucleic acid which typically specifies three or more proteins. All viruses are obligate parasites that depend on the cellular machinery of their hosts to reproduce. Viruses are not active outside of their hosts, and this has led some people to suggest that they are not alive. All types of living organisms including animals, plants, fungi, and bacteria are hosts for viruses, but most viruses infect only one type of host. Viruses cause many important plant diseases and are responsible for losses in crop yield and quality in all parts of the world.

### Basic Biology

Viruses represent not just another group of pathogens, but rather a fundamentally different form of life. Unlike all other living organisms, viruses are non-cellular. In contrast to cells, which multiply by dividing into daughter cells, viruses assemble from pools of their structural components. Mature virus particles are dormant; they come alive and reproduce only inside infected cells. In other words, viruses are obligate parasites that cannot be cultivated using any growth media suitable for bacterial, fungal, plant or animal cell types. All viruses lack protein-synthesizing and energy-producing apparatuses. As a rule, virus particles are immobile outside the infected host; they rely on the aid of other organisms or the environment for their dissemination.

### Morphology

There is a simple structural principle that applies to virtually all viruses in their mature form. Virus particles (virions) are composed of two principal parts, the genome that is made of nucleic acid, and a protective shell that is made of protein. In addition, some virus particles are enveloped by an outer membrane containing lipids and proteins (lipoprotein membrane). The protein shells of plant viruses (capsids) are assembled in accord with one of the two fundamental types of symmetry. The first type of virion is helical (roughly elongated). The elongated viruses come in two major variants, rigid rods and flexuous filaments. In both of these variants, the nucleic acid is highly ordered: it assumes the same helical conformation as the proteinaceous capsid. The second type of virus particle is icosahedral (roughly spherical); the variations of this basic shape include bacilliform virions and twin virions composed of two joined incomplete icosahedra. In the icosahedral virions, the genomic nucleic acid forms a partially ordered ball inside the proteinaceous capsid. The icosahedral and elongated virions alike can self-assemble in a test tube if the nucleic acid and protein subunits are incubated under proper conditions.

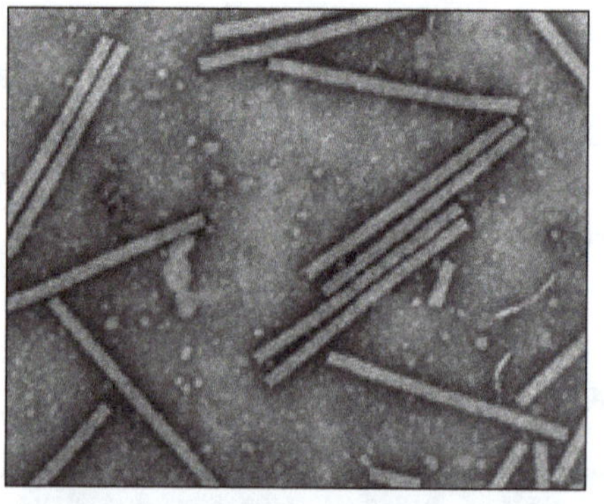

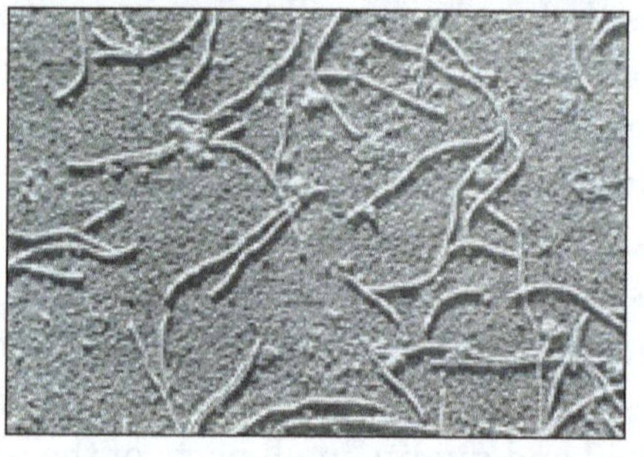

Viruses are the smallest among all known organisms. The typical diameter of a spherical plant virus is ~30 nm. The rigid, rod-shaped TMV particle is 300 x 18 nm and consists of an RNA genome of about 6,400 nucleotides encapsulated by 2,130 copies of the TMV coat protein. Some of the filamentous viruses reach the length of ~2000 nm or 2 µm. For comparison, the typical size of a leaf mesophyll cell is ~50 µm.

## Reproduction

As in other organisms, the information for virus reproduction is contained within its genome.

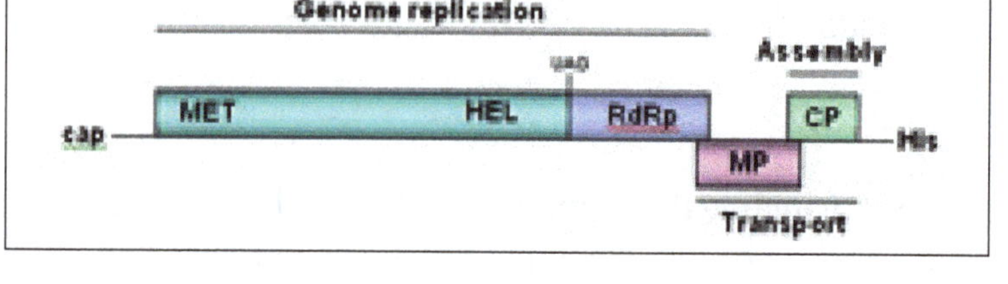

Although the genetic material for most organisms is double-stranded (ds) DNA, only a minority of plant viruses possess dsDNA genomes. Some of the plant viruses have genomes that are composed of single-stranded (ss) DNA. However, the majority of plant viruses do not use DNA at all. Instead, the genomes of nearly all plant viruses are made of RNA. Most of these genomes are composed of ssRNA that is the same (positive-sense) polarity as the messenger RNAs of the cell. Some of the RNA viruses use ssRNAs of negative polarity, and yet others have genomes made of dsRNA. Due to this enormous variation in the very nature of the genetic material of viruses, the reproductive cycles and life styles of different viruses are often very distinct from each other.

Since plant viruses are obligate, biotrophic parasites, their life cycles start by penetration of the virion into the cell. Plant viruses are unable to penetrate the plant cuticle and cell wall. It is believed that the virion enters the cytoplasm of the cell passively through wounds caused by mechanical damage to the cuticle and cell wall. The next phase of virus infection is the partial or complete removal of the coat protein shell of the virion in the cytoplasm. Next the cell mediates expression of the viral genome by providing a transcription apparatus (for DNA viruses) and a translation apparatus (for all viruses). The DNA viruses must be transported to the nucleus for transcription in order to gain access to the cell proteins required for the production of messenger RNA from viral DNA. Translation of viral RNA in the cytoplasm produces viral proteins that are required for completion of the virus life cycle.

All viruses must direct the formation of at least three types of proteins: replication proteins that are essential for nucleic acid production, structural proteins that form the protein shell and other components contained in the virions, and movement proteins that mediate virus transport between plant cells. The viral replication proteins combine with cellular proteins to produce a complex of proteins that manufactures multiple copies of the virus genome. These newly made genomes interact with the structural proteins to form new virions. The diagram below illustrates the steps that occur within the cell during the replication of Tobacco mosaic virus.

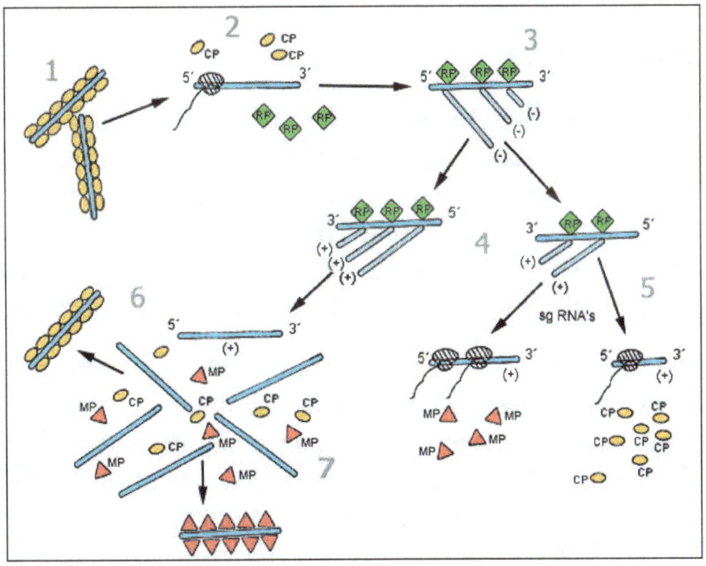


The next step in the virus reproduction cycle is movement of the virus into neighboring cells. Depending on the virus, the viral genomes or the virions are transported into neighboring cells through small channels called plasmodesmata that form connections between cells. Many viruses produce movement proteins that modify the plasmodesmata channels and facilitate viral movement into neighboring cells. The following diagram illustrates the movement of Tobacco mosaic virus from an infected cell to a neighboring cell.

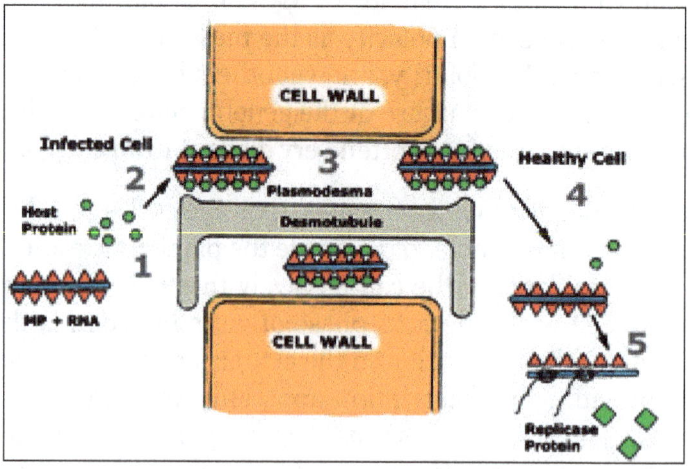


The process of cell-to-cell movement is relatively slow: it takes from one to a few hours for a virus to multiply in a cell and move to the next cell. To successfully colonize an entire plant, a virus needs to enter the vascular system of the plant. The process of systemic or long-distance transport normally precedes through the phloem sieve elements where viruses move passively with the flow of photosynthates. After quite rapid systemic spread of the virus (centimeters per hour) in the phloem, the virus moves from the phloem into surrounding cells where it reproduces and spreads by cell-to-cell movement. The time between initial infection of one or a few cells and systemic infection of the plant varies from a few days to a few weeks depending on the virus, host plant, and environmental conditions. Transmission of the virus from one plant to another completes the virus life cycle.

## Tobacco Mosaic Virus (TMV)

TMV is made up of a piece of nucleic acid (ribonucleic acid; RNA) and a surrounding protein coat. The complete virus is a submicroscopic, rigid, rod-shaped particle. Once inside the plant cell, the protein coat falls away and nucleic acid portion directs the plant cell to produce more virus nucleic acid and virus protein, disrupting the normal activity of the cell. TMV can multiply only inside a living cell but it can survive in a dormant state in dead tissue, retaining its ability to infect growing plants for years after the infected plant part died. Most other viruses die when the plant tissue dies.

The most important way that TMV can be spread from plant to plant is on workers' hands, clothing or on tools. This is called 'mechanical' transmission. When plants are handled, the tiny leaf hairs and some of the outer cells inevitably are damaged slightly and leak sap onto tools, hands, and clothing. If the sap contains TMV, it can be introduced into other plants when those come in contact with this sap. Sucking insects such as aphids do not spread TMV. Chewing insects such as grasshoppers and caterpillars occasionally spread the virus but are usually not important in spread. Vegetative propagation perpetuates TMV and other virus diseases. Cuttings taken from an infected plant usually are infected even if no symptoms are immediately exhibited by the cutting. The virus particles are found in all parts of the plant except the few cells at the tips of the growing points. Infected stock plants should be discarded immediately.

TMV can also survive outside the plant in sap that has dried on tools and other surfaces. If a TMV plant is handled and then you open a door with that hand, you have now put TMV on the door handle. The next person to open the door can pick up the TMV and spread it to any plant that they touch.

Tobacco products, particularly those containing air-cured tobacco, may carry TMV. Flue-cured tobacco, used in making cigarettes, is heated repeatedly during its processing, thereby inactivating most if not all TMV. When tobacco products are handled or kept in pockets, hands and clothing can become contaminated with TMV and be a source of virus. TMV is not spread in the smoke of burning tobacco.

Symptoms vary with the species of plant infected and the environmental conditions. In some cases environmental conditions bring out symptoms while other conditions mask or hide symptoms. Symptoms associated with TMV infections:

- Stunting.
- Mosaic pattern of light and dark green (or yellow and green) on the leaves.
- Malformation of leaves or growing points.
- Yellow streaking of leaves (especially monocots).
- Yellow spotting on leaves.
- Distinct yellowing only of veins.

Some of the above symptoms can also be caused by high temperature, insect feeding, growth regulators, herbicides, mineral deficiencies, and mineral excesses. TMV diseases cannot be diagnosed on the basis of symptoms alone.

## Managing TMV

No chemicals cure a virus-infected plant:

- Purchase virus-free plants.
- Remove all weeds since these may harbor TMV.

- Remove all crop debris from benches and the greenhouse structure.
- Set aside plants with the above symptoms and obtain a diagnosis.
- Discard infected plants.
- Disinfest tools by placing them in disinfectant for at least 10 min. Rinse thoroughly with tap water. Disinfest door handles and other greenhouse structures that may have become contaminated by wiping thoroughly with one of these materials.
- Propagate plants via seed rather than vegetatively.
- Thoroughly wash hands after handling tobacco products or TMV-infected plants.
- Do not keep tobacco products in the pockets of clothing worn into the greenhouse. Launder greenhouse work clothes regularly.

## Tomato Yellow Leaf Curl Virus (TYLCV)

Tomato yellow leaf curl virus (TYLCV) is a tomato (Solanum lycopersicum)-infecting plant virus transmitted by whitefly Bemisia tabaci. It belongs to the genus Begomovirus of the family Geminiviridae and has a single-stranded circular DNA genome of about 2.8 kb encapsidated in a twinned icosahedral virion. TYLCV-infected tomato plants show severe symptoms such as stunting, leaf curling and yellowing, which cause critical production loss in tomato cultivation. In addition to tomato, other cultivated plants including pepper (Capsicum species), common bean (Phaseolus vulgaris), cucurbit (Cucumis species) and eustoma (Eustoma grandiflora) have been reported to be TYLCV hosts. TYLCV has occurred uninterruptedly in tropical and subtropical areas. In Korea, TYLCV has been reported continuously across the country since the first outbreak in 2008.

The management of TYLCV in tomato is difficult and expensive both in cultivation under a structure and open field production. Many different approaches for controlling TYLCV disease such as removing whiteflies, killing intermediate weeds and changing cultivation season have been applied to decrease losses due to TYLCV because a single approach is not frequently effective and certain other approaches cannot be used in different agricultural environments and locations. Therefore a combination of chemical and biological control techniques for integrated pest management should be employed to reduce the population and migration of the whitefly vector and minimize or eliminate inoculum sources of TYLCV.

The transmission of most plant viruses depends on vectors such as parasitic fungi, root nematodes, mites and insects. Some viruses can also be mediated by the sap of virus-infected plants (mechanical transmission), vegetative propagation and grafting. However, these are restricted to horizontal transmission among neighboring plants. On the other hand, seed transmission can play a critical role in survival from season to season and from parent to offspring. There are three major methods of vertical transmission of plant viruses via seed contamination. A few stable viruses, such as tobacco viruses, remain only in the seed coat without embryo contamination and then are transmitted to the seedling after germination. Other methods of seed contamination correspond to virus invasion of the embryo, which is usually protected against virus invasion, either from infected maternal tissues or via infected pollen. Approximately 231 plant virus and viroid diseases have been reported to be seed-transmissible. From tomato plants, Arabis mosaic virus, Pepino mosaic virus, Tomato apical stunt viroid (TASVd), Tomato black ring virus, Tomato chlorotic dwarf viroid (TCDVd), Tomato mosaic virus and Tomato streak virus (TSV) were confirmed as transmittable by seeds. Among them, TCDVd exhibited 85.5–94.4% transmission though seeds in tomato and high transmission yields were also identified from TASVd (80%) and TSV (66%).

There have been no previous reports addressing the potential of TYLCV to be a seed-borne virus on any host plant. Begomoviruses including TYLCV have been believed to be inoculated by only whitefly-mediated transmission and artificial inoculation with an infectious clone. Abutilon mosaic virus and Beet curly top virus affiliated with the family Geminiviridae were listed as seed-borne viruses, but they were considered to be erroneous reports with nomenclature problems. Therefore, geminiviruses were thought to spread only via insect vectors in nature. Recently, however, Sweet potato leaf curl virus (SPLCV) belonging to the genus Begomovirus was first reported as a seed-transmissible virus in sweet potatoes. More than 70% of the seeds harvested from SPLCV-infected sweet potato plants were infected by SPLCV. The transmission rate of SPLCV from seeds to seedlings was up to 15%. Southern blot hybridization showing SPLCV-specific single- and double-stranded DNAs in seedlings germinated from SPLCV-infected seeds confirmed SPLCV seed transmission. Therefore, the possibility of TYLCV seed transmission has been reinvestigated based on the seed transmission results of SPLCV as another begomovirus in the family Geminiviridae.

Seed-transmissible viruses can be an initial source of inoculum for vector-mediated transmission and viruses can be disseminated over a long range and perpetuated in seeds. Therefore, understanding the possibility of TYLCV seed transmission is important for plant protection against infection and the spread of TYLCV. The careful investigation of TYLCV transmission through seeds can provide answers to unresolved problems about sporadic TYLCV occurrence in cultivation under structures or in bio nurseries without outbreak of whiteflies. In this study, TYLCV detection was performed in floral tissues, seeds and their young seedlings harvested from TYLCV-infected plants inoculated by whitefly-mediated infection and agro-inoculation to evaluate the possibility of TYLCV seed transmission in tomato plants. In addition, we also investigated whether seedlings germinated from TYLCV-infected seeds could act as TYLCV reservoirs for whitefly transmission.

## Cucumber Mosaic Virus (CMV)

Cucumber mosaic virus (CMV) was first described in detail in 1916 on cucumber and other cucurbits, but is now known to occur worldwide in both temperate and tropical climates, affecting many agricultural and horticultural crops. Development of genetic resistance to CMV in many vegetables has made a valuable contribution for disease management of this important virus disease.

### Symptoms and Signs

#### Symptoms

CMV infects 1200 species in over 100 plant families and can cause significant economic losses in many vegetable and horticultural crops. CMV causes a systemic infection in most host plants, but may remain symptomless in some crops like alfalfa. Symptoms of cucumber mosaic can vary greatly depending on the crop infected and the age of the plant when infection occurs.

Cucurbits: Almost all cucurbits are susceptible to CMV, with symptoms varying in severity. Severe epinasty, downward bending of the petiole and leaf surface along with leaf reduction, is common in early season infection of summer squash.

Plants infected early in the season are severely stunted and leaves are malformed, and fruit are unmarketable because of pronounced rugosity (roughness) on the fruit surface, as shown on the infected zucchini plant and fruit in Figure below. Infection of vining crops, such as muskmelon, show severely stunted growing tips, and although fruit may not show symptoms they are of poor quality. If the yellow squash variety grown lacks the precocious gene, color breaking will occur on the fruit, causing the fruit to show green blotchy patterns, but these symptoms are absent in yellow squash varieties with the precocious gene. Color breaking on fruit of varieties without the precocious gene will also occur with Watermelon mosaic potyvirus (WMV) infection; however this protection does not hold true for Papaya ring spot potyvirus or Zucchini yellow mosaic potyvirus, where both foliage and fruit of yellow squash are severely affected. Pumpkin is another cucurbit, that when infected at any early stage, will express severe foliar mosaic and the fruit will show a mosaic pattern and would be unmarketable.

- Pepper: Foliar symptoms of pepper plants vary with stage of infection. The initial flush of systemic symptoms typically includes a chlorosis of young leaves that may occur over the basal portion of the leaf or over the entire leaf. Oak leaf and ringspot patterns may develop on these leaves as the plant ages. As new leaves emerge, these leaves develop a chlorotic mosaic pattern that tends to encompass the entire leaf. Leaves that develop subsequent to those expressing the chlorosis and chlorotic mosaic symptoms may have varied degrees of deformation including sunken interveinal lamina with protruding primary veins. These leaves also have a dull light green appearance as opposed to the dark green, rather shiny leaves of healthy pepper plants. These symptom patterns vary in severity depending on the age of the plant at the time of infection with more severe symptoms typically resulting when plants are young at the time of infection. CMV infected plants also tend to be stunted with those plants infected early in development possibly expressing severe stunting, whereas plants infected at later stages of development may have little, if any, stunting. Pepper fruit may develop ringspotting and roughness leading to unmarketable fruit.

- Spinach: CMV infection of spinach is often referred to as spinach blight. The symptoms can vary depending upon the variety, plant age when infected, temperature, and virus strain. Typical symptoms include leaf chlorosis, which can progress to cause severe blighting of the growing point and eventual plant death. In addition to chlorotic mottle, leaves can show narrowing, wrinkling with vein distortion, and inward leaf roll.

- Lettuce: Symptoms of CMV infection of lettuce consist of leaf mottling, severe roughness of the leaf and occasional necrosis within the leaf tissue. Plants are usually stunted if infected at an early stage of development.

- Celery: CMV infection of celery is referred to as southern mosaic virus in the older literature following its description in southern production areas. Initially leaves will develop veinclearing and mosaic, and later these same leaves may show yellowing and veinal necrosis. The petioles of these plants show elongated, brown to translucent, sunken beige colored lesions, making the celery stalk unmarketable. Symptoms may also be transient, such that later in the season the symptoms may be limited to a dull cast, while under cool growing conditions the obvious symptoms may be muted.

- Tomato: Tomato plants infected with CMV in the early stages are yellow, bushy and considerably stunted. The leaves may show a mottle similar to that caused by Tomato mosaic virus (ToMV). The most characteristic symptom of CMV is filiformity or shoestring-like leaf blades.

"Shoe stringing" of CMV infected tomato.

This could be confused with a symptom of ToMV known as "fern-leaf" in which the blade of the leaflet is not as completely suppressed as in a "shoestring" leaflet, but is abnormally long and narrow. The symptoms caused by CMV can be transitory, with the bottom leaves or newly developed top leaves showing severe symptoms, while the middle leaves may appear almost normal. Severely affected plants produce few fruit, which are usually small, often mottled or necrotic, with delayed maturity.

CMV symptoms in tomato have two distinct forms depending on growing conditions. In the greenhouse: CMV-infected tomato plants grown under greenhouse conditions may express a general chlorosis followed by an easily observed mosaic symptom on young, systemically infected leaves. These symptoms tend to be somewhat transient as newly emerging leaves may be smaller than normal and show varying degrees of deformation such as shoe stringing. Internode elongation is reduced leading to overall stunting but a bushy growth habit may not occur.

In the field: The occurrence of mosaic symptoms on young, systemically-infected leaves typically includes initial chlorosis and mosaic. If plants are young at the time of infection, the mosaic symptoms are rather transient due to severe leaf deformation, and in particular, an upward cupping or rolling of leaves. If plants are somewhat more mature at the time of infection, mosaic symptoms may be more obvious, in part, because such leaves are not masked by severe leaf deformation and cupped or rolled leaves. Tomato plants that are young at the time of infection remain small, sometimes severely stunted, with a bushy appearance.

Bean: Symptoms of CMV infection in bean consist of leaf curl, green mottle and blistering, and a zipperlike roughness along the main veins involving only a few leaves. Foliar symptoms are most obvious and pod infection and loss is greatest when plants are infected before bloom. Early infected plants may yield no or few pods because CMV caused flower abortion and abnormal development. The pods are mostly curved, mottled and reduced in size. Plants may recover and resume normal growth with limited yield loss if plants are infected after bloom. Because symptoms can vary among varieties and can be confused with those caused by Bean common mosaic virus, it is best not to rely upon symptoms to diagnosis the particular virus involved. Rather the serological technique called ELISA (enzyme-linked immunosorbent assay) should be used to confirm if the plant is infected with CMV or another virus.

## Pathogen Biology

CMV is the type species of the genus Cucumovirus in the family Bromoviridae. It consists of three spherical particles, each approximately 28 nm in diameter. The CMV genome consists of three single-stranded, messenger-sense RNA molecules, designated RNA 1 (~3,350 nucleotides), RNA 2 (~3,050 nucleotides) and RNA 3 (~2,200 nucleotides). Each RNA molecule is enclosed within a protective protein coat with each being a distinct single spherical-shaped particle. Thus, a mature CMV consists of three spherical particles, one particle containing RNA 1, another containing RNA 2 and the third containing RNA 3. The RNA 3 particle may contain a fourth RNA strand, referred to as RNA 4 (~1,030 nucleotides), which encodes the coat protein gene and from which the CMV coat protein is produced. This type of translational strategy, referred to as subgenomic RNA, consists of a separate strand of RNA produced during replication. While CMV RNA 3 contains the coat protein gene, that gene is only translated to produce the coat protein from its subgenomic strand. The general replication/translation scheme for RNAs 1-3 is shown in Figure below.

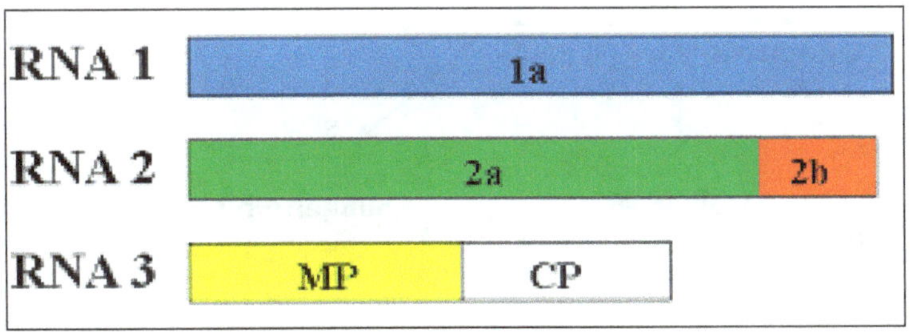


The three genomic RNAs of CMV.

RNA 1 encodes a single protein, referred to as 1a, shown to be necessary for replication of the viral genome. The 1a protein has two functional domains: the N-terminal and C-terminal domains. The N-terminal domain has methyltransferase activity for adding a cap structure to the 5'-terminus of genomic and subgenomic RNAs, while the C-terminal domain of the 1a protein is a putative helicase which functions to "unwind" the double stranded RNA that develops during viral replication.

RNA 3 encodes two proteins, referred to as 3a and coat protein. The 3a protein (MP or movement protein) is essential for virus cell-to-cell movement and mutations in this protein identified host-specific differences in movement efficiency. The 3a protein was also shown to be associated with intra-plant long distance movement. The coat protein is expressed from a sub-genomic RNA species referred to as RNA 4. The coat protein is the only protein associated with virus particles and is the sole determinant for transmission by aphid vectors. The coat protein is involved in virus cell-to-cell movement, although suggested to have an indirect role in this movement process, and is directly involved in intraplant long distance movement. It should be noted that mutations in 1a, 2a and 2b genes affected virus movement in some hosts.

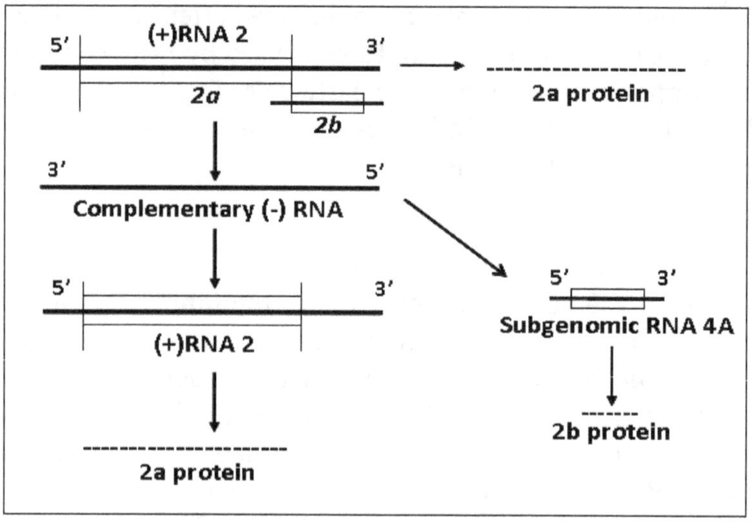


Replication of RNA 2.

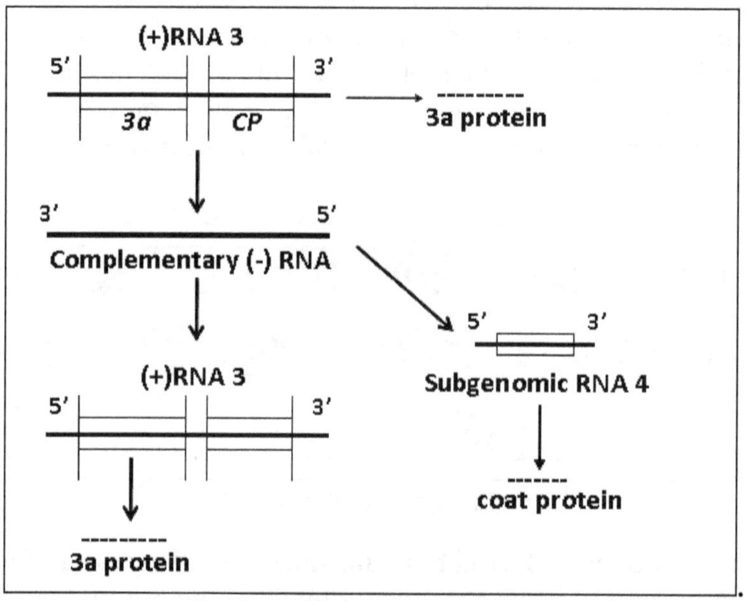


Replication of RNA 3.

In addition to the CMV genomic and sub-genomic RNA species, some strains of CMV support a satellite RNA (designated RNA 5 or satRNA). The satRNA is a single-stranded molecule of approximately 332 to 342 nucleotides in length and is completely dependent on CMV for its replication. Moreover, satRNA is encapsidated in CMV particles, which allows spread from plant to plant along with CMV, by aphid vectors. The satRNA, however, does not provide CMV (i.e., the helper virus) with any essential function. The occurrence of satRNA with CMV may have no effect on CMV-induced symptoms or may intensify symptoms by production of systemic chlorosis or necrosis or may actually attenuate symptoms typically induced by CMV. The effect on disease expression is dependent on interaction of the satRNA species, the isolate or strain of CMV and the host plant species and variety. For example, a severe lethal necrosis in tomato occurred due to one CMV-satRNA interaction that affected thousands of acres of processing tomatoes in southern Italy.

CMV replicates in the cytoplasm following a characteristic strategy used by single-stranded messenger-sense viral genomes, whereby viral genomic RNA molecules serve as templates for generation of non-translated minus-sense RNA molecules complementary in sequence to the respective genomic RNA species. Genomic and sub-genomic RNAs are generated from the respective complementary minus-sense RNA species. Proteins are translated from specific viral genes in the cytoplasm. Cell-to-cell movement occurs through plasmodesmata and intra-plant long distance movement occurs through the phloem.

CMV strains are divided into two subgroups, designated subgroups I and II that are distinguished by serological relationships and sequence analysis. Subgroup I strains are further divided into IA and IB, based on differences in pathogenicity in cowpea (Vigna unguiculata), whereby IA strains induce systemic mosaic symptoms and IB strains induce necrotic local lesions on inoculated leaves. Some CMV strains are host specific, infecting certain hosts in the same family like the legume strain of CMV. CMV is serologically related to Tomato aspermy virus and Peanut stunt virus.

CMV-satRNA interaction.

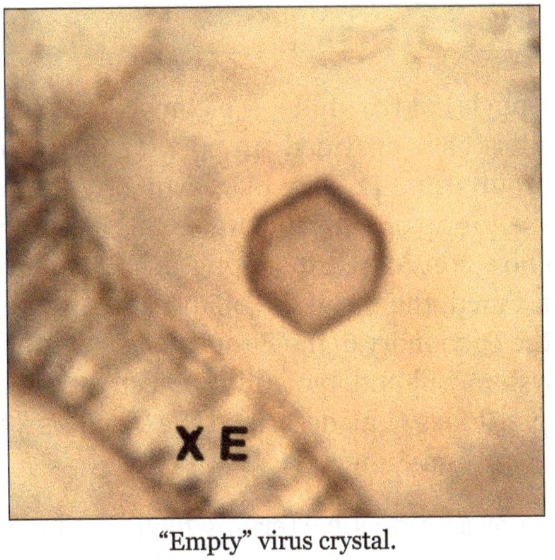


"Empty" virus crystal.

'Empty' virus crystals lacking virions located in a vacuole may appear as a neat hexagonal outline, but more often appear as 'empty' circles. Such 'empty' crystals are diagnostic for CMV, but not for cucumoviruses in general.

## Plant Pathogenic Fungi

A fungus is a eukaryote that digests food externally and absorbs nutrients directly through its cell walls. Most fungi reproduce by spores and have a body (thallus) composed of microscopic tubular cells called hyphae. Fungi are heterotrophs and, like animals, obtain their carbon and energy from other organisms. Some fungi obtain their nutrients from a living host (plant or animal) and are called biotrophs; others obtain their nutrients from dead plants or animals and are called saprotrophs (saprophytes, saprobes). Some fungi infect a living host, but kill host cells in order to obtain their nutrients; these are called necrotrophs.

Fungi were once considered to be primitive members of the plant kingdom, just slightly more advanced than bacteria. We now know that fungi are not primitive at all. In fact, recent taxonomic treatments such as the Tree of Life Project show that fungi and animals both belong to the group Opisthokonta. Fungi may not be our next of kin, but they are more closely related to animals than they are to plants. We also recognize that organisms traditionally studied as "fungi" belong to three very different unrelated groups: the true fungi in Kingdom Fungi (Eumycota), the Oomycetes, and the slime molds.

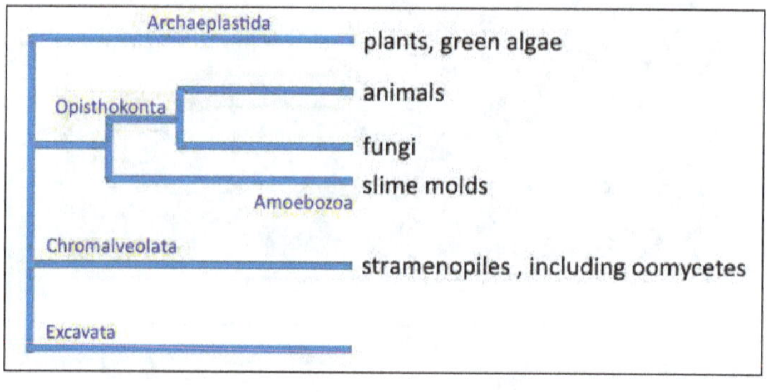


There are four main groups (phyla) of true fungi—Ascomycota, Basidiomycota, Chytridiomycota and Zygomycota. Recent studies have provided support for the recognition of additional phyla, such as Glomeromycota, a group of fungi once placed in Zygomycota that form an association with the roots of most plants. A group of parasitic organisms called Microsporidia that live inside the cells of animals are also now considered to belong in the fungal kingdom. A comprehensive classification of the Kingdom Fungi, the result of collaboration among many fungal taxonomists. This classification is used in the Dictionary of the Fungi and other fungal references and databases. However, the classification system will undergo additional changes as scientists use new methods to study the fungi. For example, Jones et al. described the "cryptomycota," a potentially new phylum of organisms within the Kingdom Fungi.

Fungi are an ancient group—not as old as bacteria, which fossil evidence suggests may be 3. 5 billion years old—but the earliest fungal fossils are from the Ordovician, 460 to 455 million years old. Based on fossil evidence, the earliest vascular land plants didn't appear until approximately

425 million years ago, and some scientists believe that fungi may have played an essential role in the colonization of land by these early plants. Mushrooms exquisitely preserved in amber from the Late Cretaceous (94 million years ago) tell us that there were mushroom-forming fungi remarkably similar to those that exist today when dinosaurs were roaming the planet. However, the fungal fossil record is incomplete and provides only a minimum time estimate for when different groups of fungi evolved. Molecular data suggest that fungi are much older than indicated by the fossil record, and may have arisen more than one billion years ago.

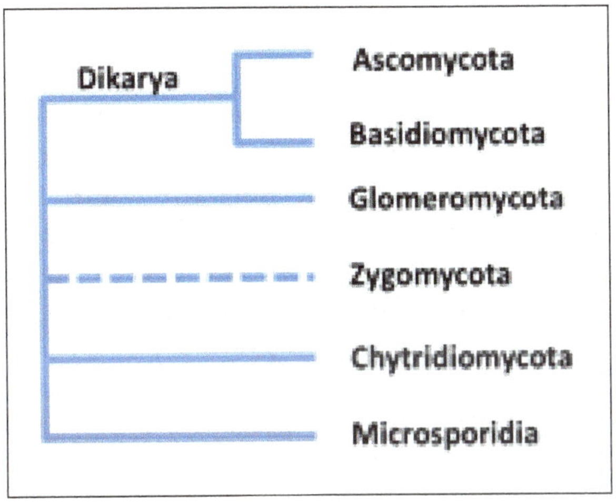


## Diversity of Fungi

No one knows for sure how many species of fungi there are on our planet at this point in time, but what is known is that at least 99,000 species of fungi have been described, and new species are described at the rate of approximately 1200 per year. A conservative estimate of the total number of fungal species thought to exist is 1. 5 million. To come up with this figure, Hawksworth estimated the known numbers of plant and fungal species from countries in which both plants and fungi have been well-studied—Great Britain and Ireland, in this case—and determined there were six fungal species for every native plant species. The total number of plant species worldwide is approximately 250,000, and if the ratio of fungi to plants in Great Britain is typical of what occurs elsewhere, there should be at least 1. 5 million species of fungi.

If 1. 5 million fungal species is a reasonable estimate, the vast majority of all extant fungi are yet to be named. Assuming a relatively constant rate at which new species are described, it will take more than 1100 years to catalog and describe all remaining fungi. However, many of these fungi are likely to become extinct before they are ever discovered given current rates of habitat and host loss. For example, up to 2% of tropical forests are destroyed globally each year. These habitats are exceedingly rich in fungal species. For example, 15-25% of fungi collected in short-term studies in the tropics are new species. Callan and Carris estimated that an 110,000 ha neotropical forest, such as in Costa Rica, could contain over 81,000 different species of plant parasitic fungi—almost as many as all the known species of fungi! Consider that this estimate was based only on plant parasitic fungi, and did not take into account other ecological groups of fungi such as saprotrophs.

Fungi are involved in a wide range of activities—some fungi are decomposers, parasites or pathogens of other organisms and others are beneficial partners in symbiosis with animals, plants or algae. Let's take a brief look at these various ecological groups.

## Fungi Associated with Animals

Fungi have the ability to grow on and in both invertebrate and vertebrate animals. Many fungi can attack insects and nematodes, for example, and may play an important role in keeping populations of these animals under control. Insect-attacking fungi, called "entomopathogens," include a wide range of fungi in phyla Ascomycota, Zygomycota and Chytridiomycota. Some of the best-known and most spectacular entomopathogens belong in the Ascomycota genus Ophiocordyceps and related genera. These fungi infect and consume insects such as caterpillars and ants, and then form conspicuous stromata that emerge from their victim's body in a most dramatic manner. These fungi can also alter the insect's behavior. "Zombie-ant" fungi from Brazil infect insect brains, directing the victim to climb up plants and bite into the plant tissue in a "death grip. Paradoxically, humans have been using one of these entomopathogens, Ophiocordyceps sinensis, for thousands of years to treat a wide range of ailments. This fungus is an important component of traditional Asian and is commonly called "winter worm, summer grass."

Entomopathogens such as Beauveria bassiana are so effective in killing insects that they are used as biological control agents for insect pests. Colony collapse disorder of honeybees has been associated with co-infection by a virus and a microsporidian fungus, Nosema ceranae. One group of fungi called Entomophthorales ("insect killers") includes a number of highly specialized entomopathogens. A common example is Entomophthora musae, which is often observed forming a ring of white spores discharged around the body of a parasitized fly on panes of glass.

Some fungi are specialized parasites of nematodes, rotifers, and other microscopic animals in the soil. A common nematode predator is Arthrobotrys oligospora, a fungus that has evolved sticky networks of hyphae for trapping nematodes. Once the nematode is immobilized, the fungus invades and consumes its body.

Fortunately, there are relatively few fungal pathogens of vertebrates—only 200-300 species—but some of these fungi can have devastating impacts. Consider the well-publicized frog killer, Batrachochytrium dendrobatidis, a member of phylum. This fungus wasn't even known to scientists until 1996, when it was discovered associated with frogs that had died from a mysterious skin disease at the Smithsonian National Zoological Park in Washington, D. C. The fungus doesn't invade the frog's body, but it is lethal, possibly because it disrupts electrolyte balance leading to cardiac arrest—infected frogs appear to die of a heart attack! The frog chytrid is implicated in the widespread decline of frog populations around the world. Fortunately, this is the only chytrid known to parasitize a vertebrate animal and it appears to infect only amphibians.

Another devastating parasite of animals is Geomyces destructans, a cold-loving fungus that causes 'white-nose syndrome' in bats. This fungus colonizes the skin on the muzzles, ears and wing membranes of some types of bats, and infected bats exhibit unusual behavior. The bat fungus is associated with declines in bat populations in the northeastern U. S. and has many wildlife biologists concerned. As of 2011, white-nose syndrome had been confirmed in 16 states and four Canadian provinces.

In humans, there are several different types of fungal infections, or "mycoses." The most common are caused by dermatophytes, fungi that colonize dead keratinized tissue including skin, finger-, and toenails. Dermatophytes cause superficial infections such as 'ringworm' that are unsightly and difficult to treat, but rarely serious. Some fungi are members of the resident micro-flora in healthy people, but become pathogenic in people with predisposing conditions. For example, Candida species cause annoying yeast infections in the mucosal tissues of many healthy people, but can also cause diseases collectively called candidiasis in babies and immune-compromised individuals. Another group of fungi are inhaled as spores and initiate infection through the lungs. These fungi include Coccidioides immitis (coccidioidomycosis, commonly known as valley fever), and Histoplasma capsulatum (histoplasmosis). Opportunistic fungal pathogens are normally not associated with humans and other animals, but can cause serious infections in weakened or healthy individuals when inhaled or implanted in wounds. Aspergillus fumigatus, one of the most important of these opportunists, produces small, airborne spores that are frequently inhaled; in some individuals the fungus starts growing invasively, causing a disease known as aspergillosis, especially in immune-compromised individuals.

A remarkable discovery was that Pneumocystis carinii, the organism causing pneumonia-like symptoms in immune-compromised patients, is a fungus and not a protozoan as had been thought for decades. Why was this pathogen classified as a protozoan? It does not respond to the common drugs used to treat fungal infections, but does respond to anti-protozoan drugs. This unusual fungus emerged as one of the leading causes of death in AIDS patients in the late twentieth century.

## Fungi and Plants

The association of fungi and plants is ancient and involves many different fungi. Fungi are an important group of plant pathogens—most plant diseases are caused by fungi—but fewer than 10% of all known fungi can colonize living plants. Plant pathogenic fungi represent a relatively small subset of those fungi that are associated with plants. Most fungi are decomposers, utilizing the remains of plants and other organisms as their food source. Other types of associations include the role of fungi as decomposers, as beneficial symbionts, and as cryptic plant colonizers called endophytes.

Most fungi are associated with plants as saprotrophs and decomposers. These fungi break down organic matter of all kinds, including wood and other types of plant material. Wood is composed primarily of cellulose, hemicellulose, and lignin. Lignin is a complex polymer that is highly resistant to degradation, and it encrusts the more readily degradable cellulose and hemicellulose. Fungi are among the few organisms that can effectively break down wood, and fall into two main types—brown and white rot fungi. Brown rot fungi selectively degrade the cellulose and hemicellulose in wood, leaving behind the more recalcitrant lignin. The decayed wood is brown in color and tends to form cubical cracks due to the brittle nature of the remaining lignin. Only 10% of the wood decay fungi cause brown rot, and most of these fungi (80%) occur on conifer wood. Brown rot residues make up 'humus' in temperate forest soils and are important for mycorrhizal formation, moisture retention, and for sequestering carbon. Brown rot residues are highly resistant to decomposition and can remain in the soil for up to 300 years. White rot fungi are more common than brown rot fungi; these fungi degrade cellulose, hemicellulose, and lignin at approximately equal rates. The decayed wood is pale in color, light in weight, and has a stringy texture. White rot fungi are the only organisms that can completely degrade lignin. Lignin is one of the most abundant organic polymers, accounting for 30% of the organic carbon on the planet—only cellulose is more abundant.

An important group of fungi associated with plants is mycorrhizal fungi. Mycorrhiza means 'fungus root', and it refers to a mutually beneficial association (a type of symbiosis) between fungi and plant roots. There are seven major types of mycorrhizal associations, the most common of which is the arbuscular mycorrhizae, involving members of phylum Glomeromycota associated with roots of most major groups of plants. The vast majority (>80%) of vascular plants form mycorrhizae.

Another common type of association is ectomycorrhizae formed between forest trees and members of phyla Basidiomycota and Ascomycota. In this association, the fungus forms hyphae around host root cortical cells—the "Hartig net"— and a sheath of hyphae around the host roots called a "mantle. "Many of the ectomycorrhizal fungi are mushroom-forming species including highly prized edibles such as chanterelles (Cantharellus cibarius and related species), boletes (Boletus edulis and related species), and matsutake (Tricholoma magnivelare). A valuable group of ectomycorrhizal fungi are truffles, members of phylum Ascomycota that form underground fruiting bodies. The French Périgord truffle, Tuber melanospora, and the Italian white truffle, Tuber magnatum, can bring phenomenal prices; for example, a 1. 5-kg Italian white truffle sold for $330,000 at an auction in 2007.

Lichens are examples of a symbiotic association involving a fungus and green algae or less frequently Cyanobacteria. The lichen thallus is composed mostly of fungal hyphae, usually with the alga or cyanobacterium confined to discrete areas of the thallus. In lichens, reproductive structures of the fungus are often conspicuous, for example disc- or cup-like structures called apothecia. The fungus obtains carbohydrates produced by photosynthesis from the algae or cyanobacteria, and in return provides its partner(s) with protection from desiccation and ultraviolet light. Lichens grow in a wide range of habitats on nearly every continent. Think about an inhospitable place, and there's probably lichen that grows there—on bare rocks, sidewalks, grave stones, the exoskeletons of some insects, and even on cars that remain for a long time in one place.

Some fungi are hidden inside their plant hosts; these are endophytes, defined by their presence inside asymptomatic plants. All plants in natural ecosystems probably have some type of symbiotic association with endophytic fungi. Endophytic fungi have been shown to confer stress tolerance to their host plant, for example, to disease, herbivory, drought, heat, salt and metals. The clavicipitaceous endophytes in the genus Neotyphodium (phylum Ascomycota) are among the best studied. These fungi produce alkaloid compounds that protect the grass host from insects that would otherwise feed on them; endophyte-infected turfgrass seed is sold commercially for seeding lawns and other types of grassy recreational areas. Unfortunately, livestock such as sheep, cattle, llamas and horses also are negatively affected by toxins produced by endophytes when they eat infected grass. 'Ryegrass staggers' occurs when animals graze on perennial ryegrass (Lolium perenne) that is colonized by Neotyphodium lolii. Afflicted animals develop symptoms including tremors and jerky or uncoordinated movements.

Let's now consider the role of fungi as plant pathogens. There are thousands of species of plant pathogenic fungi that collectively are responsible for 70% of all known plant diseases. Plant pathogenic fungi are parasites, but not all plant parasitic fungi are pathogens. What is the difference between a parasite and a pathogen? Plant parasitic fungi obtain nutrients from a living plant host, but the plant host doesn't necessarily exhibit any symptoms. In this sense, endophytic fungi are plant parasites because they live in intimate association with plants and depend on them for nutrition. Plant pathogenic fungi are parasites and cause disease characterized by symptoms.

Biotrophic fungal pathogens obtain nutrients from living host tissues, often via specialized cells called haustoria that form inside host cells. Necrotrophic pathogens obtain nutrients from dead host tissue, which they kill through the production of toxins or enzymes. Most biotrophic fungi have fairly narrow host ranges—they are specialized on a limited number of plant hosts. Necrotrophic fungi can be either generalist, growing on a wide range of host species, or specialized on a restricted range of hosts. Some plant pathogenic fungi change the way that their hosts grow, either by affecting the level of growth regulators produced by the plant, or by producing growth regulators themselves. Examples of changes in plant growth caused by plant pathogenic fungi include cankers, galls, witches' broom, leaf curl and stunting.

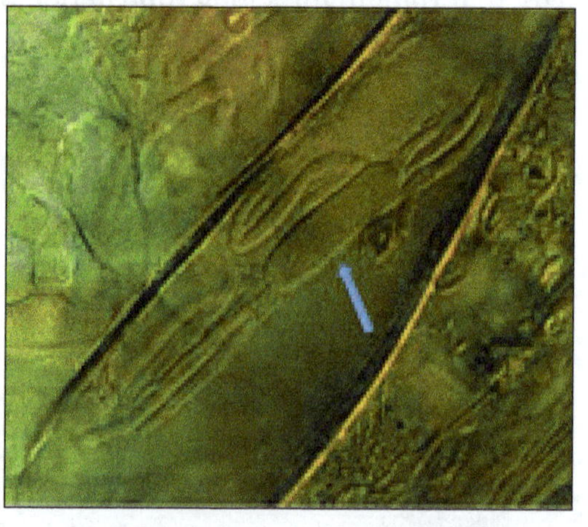

We can further divide plant pathogenic fungi by the stage of the plant host that is attacked, for example, seeds, seedlings, or adult plants, and by what part of the plant is affected—roots, leaves, shoots, stems, woody tissues, fruits or flowers. A group of fungi including species of Fusarium, Rhizoctonia and Sclerotium cause seed rot and infect plants at the seedling stage. These pathogens can attack a wide range of plants. Often, seedling pathogens cause damping-off symptoms because they occur in wet soils.

Many of the same fungi that kill seedlings can also infect the roots of mature plants and cause root and crown rot diseases. Infection often occurs through wounds, and results in lesions or death of part or all of the root system and crown. Some common root rots of trees are caused by members of phylum Basidiomycota in the genera Armillaria and Heterobasidion. Armillaria spp. produce shoe-string-like bundles of hyphae called rhizomorphs that allow the fungus to grow from one tree to another. Species of Heterobasidion survive as saprotrophs in dead tree stumps and roots, but can also infect living hosts through root contact. These fungi cause decay in the roots and crown; infected trees become weakened and die, or may blow over in high winds. Wood rot fungi, most of which are also members of Basidiomycota, infect trees through wounds, branch stubs and roots, and decay the inner heartwood of living trees. Extensive decay weakens the tree, and reduces the quality of wood in trees harvested for timber.

Vascular wilt pathogens kill their host by infecting through the roots or through wounds and growing into the xylem, where they produce small spores that get carried upward until they are trapped at the perforated ends of the xylem vessels. The spores germinate and grow through the pores. The fungus is transported throughout the plant in this manner. The first symptom of vascular wilt is a loss of turgidity in the plant leaves, often on one side of the plant or a single branch. If the stems of infected plants are cut open, vascular discoloration is evident. Among the important vascular wilt fungi are Fusarium oxysporum, Verticillium albo-atrum and V. dahliae.

One of the most famous vascular wilts is Panama disease of bananas, caused by Fusarium oxysporum forma specialis (f. sp.) cubense. This fungus nearly wiped out banana production in Latin America in the early twentieth century. Most bananas that were being grown for export were a single cultivar, 'Gros Michel', which turned out to be highly susceptible to Panama disease. There is no effective method for controlling Panama disease and it rapidly spread throughout banana plantations around the world. The banana industry was saved by the discovery of the cultivar 'Cavendish' that is resistant to the strain of Panama disease that killed 'Gros Michel'. 'Cavendish' is now the banana that Americans and Europeans consume, but a new strain of the Panama disease pathogen began killing 'Cavendish' in Malaysia in 1985 and scientists are concerned that this strain will begin to spread.

Leaf spot pathogens infect through natural plant openings such as stomates or by penetrating directly through the host cuticle and epidermal cell wall. In order to penetrate directly, fungi produce hydrolytic enzymes—cutinases, cellulases, pectinases and proteases—for breaking down the host tissue. Alternatively, some fungi form specialized structures called appressoria (sing. appressorium) at the end of germ tubes. Turgor pressure builds up in the appressorium, and in combination with an infection peg, mechanical force is exerted to breach the host cell walls. Once inside the plant leaf, the fungus must obtain nutrients from the cells, and this is often accomplished by killing host cells (necrotrophs). Death of host cells is evident as an area of dead cells called a lesion.

Many leaf-spotting fungi produce toxins that kill host cells and this often produces a lesion surrounded by a yellow halo. If enough of the leaf surface is killed, or if the infected leaves drop prematurely, the plant's ability to produce photosynthates is severely impaired.

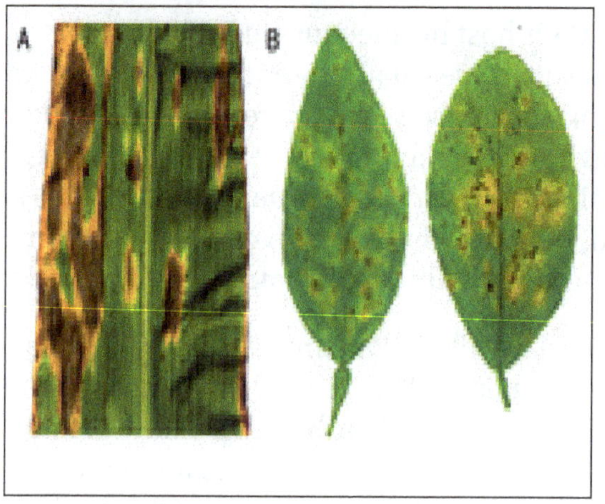


Returning to bananas, another devastating disease of this host is black leaf streak, or black Sigatoka, caused by Mycosphaerella fijiensis. Unlike the root-infecting pathogen that causes Panama disease, the black Sigatoka pathogen can be controlled by applications of a protective fungicide to banana leaves. Effective control of black Sigatoka requires multiple fungicide applications and control of this disease accounts for up to 25% of the total production cost for bananas.

American chestnut trees were once a prominent hardwood tree in the eastern U. S., but have largely been eliminated by the chestnut blight pathogen, Cryphonectria parasitica, an example of a canker-causing fungal pathogen. Cankers develop when the pathogen kills the phloem and vascular cambium in a woody host. If the canker encircles the trunk or branch of a tree, that plant part will die. The canker-causing fungus can often be identified based on the fruiting bodies that form in the canker. In contrast to cankers, galls result from abnormal growth of a plant, usually due to an increase in cell size and cell division. Although galls are often associated with insect pests, some fungal pathogens induce galls; two common examples are the black knot pathogen, Apiosporina morbosa on Prunus spp. and species of Gymnosporangium, which induce the formation of galls on their coniferous hosts.

Gymnosporangium is a type of rust fungus. Rust fungi are biotrophic pathogens—they infect, grow, and sporulate in living plant tissue. Even though biotrophs require living host tissue for their growth and reproduction, they can be devastating pathogens by reducing the photosynthetic surface and increasing water loss in the host plant. Rust fungi attack a wide range of plants, and often require two, unrelated hosts in order to complete their life cycles. Rust fungi are so-named because of the abundant orange spores that are formed on plants that are infected by these fungi; infected plants often look as though they are rusting.

One historically important rust fungus is black stem rust of wheat, a disease that was well known to the ancient Romans. Black stem rust, caused by Puccinia graminis f. sp. tritici, infects wheat and barberry (Berberis species). Since the barberry host is required for the pathogen to complete its life cycle, early control measures in the United States and Canada were aimed at eliminating this host, not the economically important host, wheat. We now know that this method of eradication was of limited success because rust spores can be carried long distances—for example, from northern Mexico to the U. S. -Canada border—by wind currents via the "Puccinia pathway.

## Morphological Characteristics of Fungi: Mycelium and Hyphae

Now let's take a closer look at fungi and the types of structures that they form. A key characteristic of fungi that has contributed to their successful exploitation of diverse ecological niches is the formation of a filamentous thallus called the mycelium. A mycelium is composed of branching, microscopic tubular cells called hyphae that grow through and across substrates or food sources, secreting enzymes that break down complex substrates into simple compounds that can be absorbed back through the cell wall. The fungal cell wall in the Kingdom Fungi is composed of chitin and glucans (in Ascomycota, Basidiomycota and Chytridiomycota) as well as chitosan and other components (in Zygomycota).

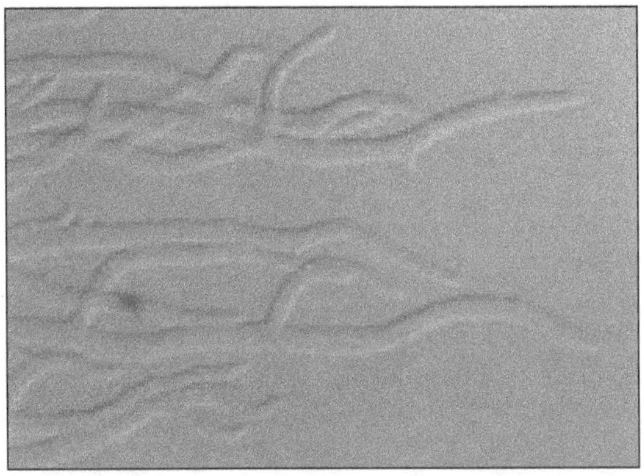

Hyphae can have cross walls called septa, or lack cross walls (non-septate; aseptate; coenocytic). The type of hyphae—septate or aseptate—is characteristic of specific groups of fungi. In fungi that form septate hyphae, there are perforations at the septa, called septal pores, which allow the movement of cytoplasm and organelles from one compartment to the next. The type and complexity of the septal pore is characteristic of specific groups of fungi.

Hyphae grow from a germinating spore or other type of propagule. Hyphae elongate almost exclusively at the tips, growing outwards from the point of establishment. As a result of apical growth, hyphae are relatively uniform in diameter, and mycelium that grows in an unimpeded manner forms a circular colony on solid substrates that support fungal growth; agar, a gelatinous material derived from seaweed, amended with different types of nutrients is commonly used to grow fungi in culture.

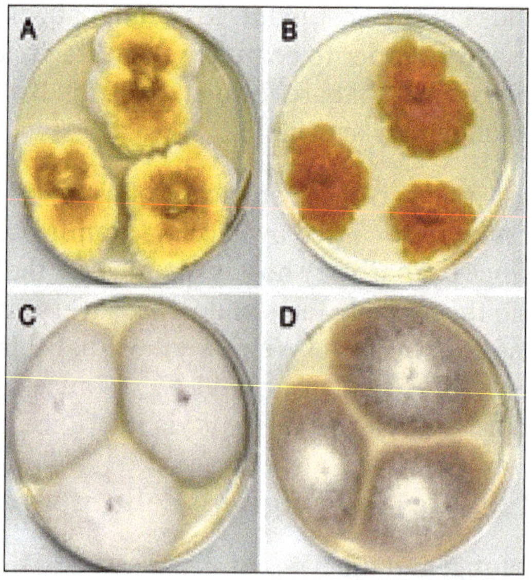


Some fungi grow exclusively or mostly as yeasts, defined as single-celled fungi that reproduce by budding or fission. In contrast to apical growth that is characteristic of hyphae, yeasts exhibit wall growth over the entire cell surface, often resulting in a nearly spherical cell. There are also fungi that can switch between mycelial growth and yeast-like growth, dependent upon the environmental conditions. The ability to grow in different forms is called dimorphism, and is exhibited by some members of phyla Ascomycota, Basidiomycota and Zygomycota.

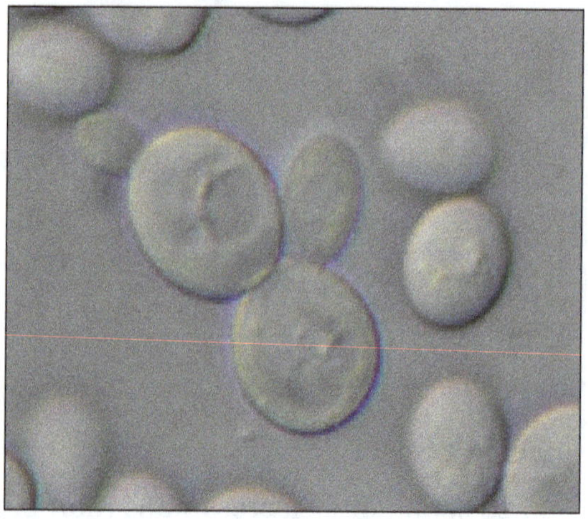

## Inside the Fungal Cell

Most of the organelles present in fungal cells are similar to those of other eukaryotes. Fungal nuclei are usually small (< 2 μm diameter), and can compress and/or stretch to move through septal pores and into developing spores. Fungi have been found to possess between 6 and 21 chromosomes coding for 6,000 to nearly 18,000 genes. Genome sizes range from 8. 5 mega base pairs (Mb) to just over 400 Mb in filamentous, making fungal genomes among the smallest of eukaryotic organisms on average—approximately 1% the size of mammalian genomes and only 1. 3 times the size of the largest known bacterial genome. Many fungi (Ascomycota) have a life cycle that is predominantly haploid, while others (Basidiomycota) have a long dikaryotic phase.

## Fungal Reproduction

Fungi frequently reproduce by the formation of spores. A spore is a survival or dispersal unit, consisting of one or a few cells, that is capable of germinating to produce a new hypha. Unlike plant seeds, fungal spores lack an embryo, but contain food reserves needed for germination. Many fungi produce more than one type of spore as part of their life cycles. Fungal spores may be formed via an asexual process involving only mitosis (mitospores), or via a sexual process involving meiosis (meiospores). The manner in which meiospores are formed reflects the evolutionary history and thus the classification for the major groups (phyla) of fungi.

Many fungi produce spores inside or upon a fruiting body. Many people are familiar with the mushroom, a type of fruiting body produced by some Basidiomycota. You may recognize other fungal fruiting bodies such as puffballs, or shelf fungi. These are examples of large, conspicuous fruiting bodies, but there is an even greater diversity of microscopic fruiting bodies produced by various fungi. What all fruiting bodies have in common is that they produce spores and provide a mechanism for dispersing those spores.

## Teleomorph and Anamorph

Many fungi are able to reproduce by both sexual and asexual processes. Sexual and asexual reproduction may require different sets of conditions (e. g., nutrients, temperature, light, moisture). In some fungi, two sexually compatible strains must conjugate (mate) in order for sexual reproduction to occur. The terms 'anamorph' and 'teleomorph' are used to convey the asexual and sexual reproduction morphological types, respectively, in a particular fungus. The concept of anamorph and teleomorph is a confusing one for many students, as we are not accustomed to thinking about organisms with such reproductive flexibility.

## Meiospores

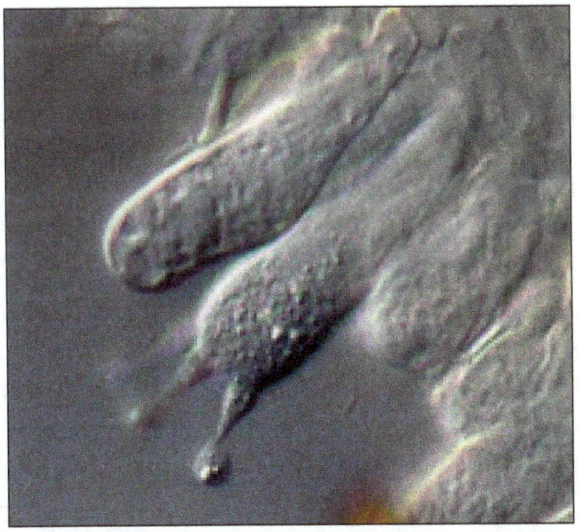

Examples of meiospores—spores that are the products of meiosis—include ascospores and basidiospores. Ascospores are formed inside a sac-like structure called an ascus. An ascus starts out as a sac of cytoplasm and nuclei and by a process called "free cell formation" a cell wall forms de novo around each nucleus and surrounding cytoplasm to form ascospores (typically eight per ascus).

Ascospores vary in size, shape, color, septation, and ornamentation among taxa. Basidiospores are formed on a basidium and are typically one-celled with one or two haploid nuclei. Basidiospores vary in size, color and ornamentation depending upon the taxonomic group.

## Mitospores

Examples of mitospores are conidia (sing. conidium), sporangiospores, and zoospores, formed by members of the phyla Ascomycota, Zygomycota, and Chytridiomycota, respectively. Another type of asexual propagule produced by fungi in several different phyla is the chlamydospore.

## Conidia

Conidia are formed from a modified hypha or a differentiated conidiogenous cell of the fungus. Conidiogenous cells can be formed singly on hyphae, on the surface of aggregated hyphal structures, or within different types of fruiting bodies. Fruiting bodies inside which conidia are formed are pycnidia and acervuli. Sporodochia and synnemata are examples of fruiting bodies on which conidia are formed. Conidia are produced primarily by Ascomycota, although some Basidiomycota are capable of producing them as well.

## Sporangiospores

Sporangiospores are asexual propagules formed inside a globose or cylindrical sporangium by a process involving cleavage of the cytoplasm. Sporangiospores are thin-walled, one-celled, hyaline or pale-colored, and are usually globose or ellipsoid in shape. One to 50,000 sporangiospores may be formed in a single sporangium. When mature, sporangiospores are released by breakdown of the sporangial wall, or the entire sporangium may be dispersed as a unit. Sporangiospores are produced by fungi in phyla Chytridiomycota and Zygomycota, as well fungal-like Oomycetes.

## Zoospores

A zoospore is a microscopic, motile propagule, approx. 2 to 14 μm long and 2 to 6 μm in diameter that lacks a cell wall and is characterized by having one or more flagella. Flagella are ~ 0. 25 μm in diameter and up to 50 μm long. Zoospores are produced by one group of true Fungi (Chytridiomycota) and by fungal-like organisms in Kingdom Straminipila and some slime. Two types of flagella are known—the whiplash flagellum, which is directed backward, and the tinsel flagellum, which is directed forward. The tinsel flagellum is only present in members of Kingdom Straminipila and does not occur in true fungi. The length of time zoospores are able to swim is determined by their endogenous energy reserves—zoospores cannot obtain food from external sources—and environmental conditions. Zoospores may exhibit chemotaxis—movement in response to a chemical gradient, e. g., root exudates. At the end of its motile phase, the zoospore undergoes a process called encystment in which it either sheds or retracts the flagella and produces a cell wall. The encysted zoospore, called a cyst, may germinate directly by the formation of a germ tube, or indirectly by the emergence of another zoospore.

Zoospores are formed inside a sac-like structure called a zoosporangium by a process involving mitosis and cytoplasmic cleavage—similar to the formation of sporangiospores in sporangia. Depending upon the taxonomic group, zoospores emerge from the zoosporangium through breakdown of the zoosporangial wall, through a preformed opening in the wall covered with a cap called an operculum that flips back, or by a gelatinous plug that dissolves.

## Chlamydospores

Chamydospores are survival propagules formed from an existing hyphal cell or a conidium that develops a thickened wall and cytoplasm packed with lipid reserves. The thickened cell walls may be pigmented or hyaline, and chlamydospores develop singly or in clusters, depending upon the fungus. Chlamydospores are passively dispersed, in most instances when the mycelium breaks down. Chlamydospores are formed by many different groups of fungi and are often found in aging cultures.

## Sclerotia

Sclerotia (sing. sclerotium) are compact aggregations of hyphae differentiated into an outer, pigmented rind, and an inner mass of hyaline cells called a medulla. Sclerotia contain food reserves, and are a type of survival propagule produced by a number of fungi in phyla Ascomycota and Basidiomycota; in some fungi, such as Rhizoctonia solani, they are the only type of propagule produced, whereas in fungi such as Claviceps purpurea and Sclerotinia sclerotiorum, they are overwintering structures that can germinate directly, or give rise to structures in which the meiospores are formed.

## Kingdom Fungi

The characteristics and diversity of the major phyla of true Fungi will be briefly described. Selected representatives of the different phyla are introduced and, in many instances, illustrated. A generalized life cycle also is presented for each phylum that illustrates when plasmogamy (cell fusion), karyogamy (nuclear fusion) and meiosis occur relative to each other, and the types of structures involved in these events.

Phylum Ascomycota is the largest group of fungi, with approximately 33,000 described species in three subphyla—Taphrinomycotina, Saccharomycotina, and Pezizomycotina. Members of this phylum reproduce sexually or meiotically via production of ascospores inside a sac-like structure called an ascus.

## Botrytis Cinerea

The pathogen Botrytis cinerea Pers. Fr. (teleomorph Botryotinia fuckeliana (de Bary) Whetzel) causes serious losses in more than 200 crop species worldwide. It is most destructive on mature or senescent tissues of dicotyledonous hosts, but it usually gains entry to such tissues at a much earlier stage in crop development and remains quiescent for a considerable period before rapidly rotting tissues when the environment is conducive and the host physiology changes. Therefore, serious damage is caused following harvest of apparently healthy crops and the subsequent transport to distant markets where the losses become evident. However, B. cinerea also causes massive losses in some field- and greenhouse-grown horticultural crops prior to harvest, or even at the seedling stage in some hosts. Some monocotyledonous hosts are also susceptible to attack by B. cinerea, and there is a group of related Botrytis species specialized to infect about a dozen monocot hosts.

B. cinerea is difficult to control because it has a variety of modes of attack, diverse hosts as inoculum sources, and it can survive as mycelia and/or conidia or for extended periods as sclerotia in crop debris. For these reasons the use of any single control measure is unlikely to succeed and a more detailed understanding of the host–pathogen interaction, the microenvironment in which the fungus operates and its microbial competitors on the host is essential. The current cost of bringing a new fungicide or biological control agent to market is so high that only major crops attract sufficient interest by agribusiness.

## Taxonomy and Genetic Variation

Phylogenetic analysis of 22 species of the genus Botrytis revealed that B. cinerea forms a small clade with three other species, all of which are specialized pathogens of dicots. Beever and Weeds reviewed the current status of B. cinerea and its teleomorph Botryotinia fuckeliana. The conidia (macroconidia) are multinucleate and the microconidia (male spermatia) are uninucleate. Shirane et al. reported 16 chromosomes at mitotic metaphase and Faretra and Grindle also found 16 chromosomes in developing asci. Apothecia of B. cinerea are rare in the field, but are more common in other Botrytis spp. Most strains are heterothallic, carrying one or other allele of the mating type locus MAT1-1 or MAT1-; however, there are also field isolates with dual mating. Sexual crossing in vitro involves incubating sclerotia of the female (sclerotial) parent for long periods at zero degrees (preconditioning) before fertilizing them with a suspension of microconidia from the spermatial parent obtained by irrigation of an ageing culture.

Genetic variation in B. cinerea populations has been studied using a variety of molecular techniques including RFLP analysis of PCR-amplified loci, PCR detection of transposable elements, RAPD fingerprinting, amplified fragment length polymorphism (AFLP) analysis, fingerprinting of repetitive sequences by microsatellite primed (MP)-PCR, PCR amplification of microsatellite loci and DNA sequencing of gene. Based on multiple gene genalogies, Fournier et al. postulated that B. cinerea is a species complex comprising two phylogenetic species, but this hypothesis has not yet been fully adopted by the Botrytis community.

## Host Range

Droby and Lichter provide a comprehensive list of post-harvest rots caused by B. cinerea; these range from grey mould on different plant organs, including flowers, fruits, leaves, shoots and soil storage organs (i.e. carrot, sweet potato), although the fungus is not regarded as a true root pathogen or one causing soil-borne diseases. Vegetables (i.e. cabbage, lettuce, broccoli, beans) and small fruit crops (grape, strawberry, raspberry, and blackberry) are most severely affected. With increasing international trade in cold-stored produce this fungus has attained great importance because it can grow effectively over long periods at just above freezing temperatures in products such as kiwifruit, apples ('dry eye rot') and pears. Similarly, the important trade in cut flowers is adversely affected by this pathogen; rose and gerbera flowers are particularly prone to damage. Culture of plants out-of-season in heated or unheated greenhouses and under plastic tunnels used increasingly to supply fruits, vegetables, herbs and flowers in northern latitudes greatly increases the risk of infection, especially in tomato, cucumber and sweet pepper.

There are important field crops that sustain serious damage due to grey mould. Most notable are the heavy losses in chickpeas and other protein-rich legumes that support millions of rural families in India, Bangladesh and Nepal. French bean (Phaseolus vulgaris) can sustain almost complete losses. Most legumes suffer attack by B. cinerea to some extent, but field bean (Vicia faba) is also damaged by chocolate spot caused by B. fabae. Blossom blight in alfalfa in Canada causes serious problems in seed production in irrigated areas. Sunflower is an important oil crop that can be infected severely. In tree nurseries, conifer seedlings are vulnerable to grey mould. With increasing interest in renewable technologies and sustainable development it is important to note also that industrial hemp (Cannabis sativa) used for fibre production is susceptible to head blight.

## Symptoms

B. cinerea is responsible for a very wide range of symptoms and these cannot easily be generalized across plant organs and tissues. Soft rots, accompanied by collapse and water-soaking of parenchyma tissues, followed by a rapid appearance of grey masses of conidia are perhaps the most typical symptoms on leaves and soft fruits. In thick-skinned fruits, such as kiwifruits, the dark water-soaking symptom is evident only after cutting. On many fruits and vegetables the infection commonly begins on attached senescent flowers and then as a soft rot it spreads to affect the adjacent developing fruit (blossom-end rot), as in courgettes (zucchini), cucumbers, French beans, strawberries and apples. On flower petals, symptoms range from minute 'pock' marks to full-scale soft rotting depending on the environmental conditions. In greenhouse-grown tomato, the greatest damage occurs on stems at pruning wounds where the fungus can rot through the entire stem. Soft rotting of mature tomato fruits occurs mainly post-harvest; an unusual 'ghost spot' symptom in unripe tomato is associated with a successful host defence, but the symptom renders fruits unmarketable.

Symptoms of infection by Botrytis cinerea. (a) Grey mould of strawberry fruit. (b) Grey mould of raspberry fruit. (c) Rose petals blemished by lesions (right) following inoculation with dry conidia and incubation at 100% RH for 48 h, compared with non-inoculated control (left). Effect of humidity on infection of rose petals by dry-inoculated conidia of Botrytis cinerea. (d) Lesions arising at nodes following infection of raspberry leaves in autumn. (e) Mummified 1-year-old blackcurrant fruits attached to stem, releasing conidia amongst newly opened flowers.

In red raspberry (Rubus idaeus), apart from the devastating grey mould on fruits, the pathogen attacks mature-to-senescent leaves causing a wedge-shaped chestnut brown lesion with yellow margin that spreads to the node on vegetative stems (primocanes) to give rise to a conspicuous pale brown fast-spreading lesion (up to 15 cm) in the primary cortex of the stem. Such infection does not enter the axillary buds because of periderm layers, but it retards development of the buds at infected nodes with the result that they fail to produce fertile lateral shoots in the following year. After winter dormancy, the stem lesions in raspberry become white and show large black sclerotia that produce masses of grey conidia in spring. In blackcurrant, symptomless infection of flower styles (detected by fluorescence microscopy) by B. cinerea leads to premature abscission of developing fruits associated with ethylene generation in a condition called 'run-off'.

Seed-borne infection has been reported in over 50 hosts; including flax, sunflower and lettuce. Seed transmission occurs in chickpeas, and in Australia it can cause total crop failure. Grey mould in this crop often begins by rotting of the herbaceous stems at ground level, with other soft-rot lesions appearing also on leaves and pods. Barnes and Shaw described the occurrence of widespread internal infection in Primula × polyantha plants grown from infected commercial seeds with symptoms of disease only appearing 3 months later at flowering. This apparent endophytic relationship with the host remains to be studied by modern microscopic tools.

## Life Cycle and Epidemiology

Sclerotia develop within dying host tissues and represent an important survival mechanism in B. cinerea, but they are very variable in size, and are not readily apparent in all susceptible crops. The melanized rind and β-glucans encasing the internal mycelium protect sclerotia from desiccation, UV radiation and microbial attack over long periods. Sclerotia commence growth in early spring in temperate regions to produce conidiophores and multinucleate conidia, serving as a primary source of inoculum within a crop. Mycelium also survives within infected dead host tissues left as crop debris and inside some seeds to serve as primary inoculum. In perennial crops, the dead leaves, flowers and mummified fruits contain masses of mycelium that can often be ideally situated within a crop canopy to produce conidia and initiate infections. The pathogen also forms microconidia from phialides abundantly in ageing cultures, which function primarily as spermatia. The sexual cycle involves the spermatization of sclerotia, leading to the production of apothecia and asci with eight binucleate ascospores. The cellular details of plasmogamy and initiation of apothecia have still not been described. Furthermore, the apothecia are unrecorded or rare in most crops attacked by B. cinerea and any conclusions about the role of the sexual cycle in the species are based mainly on molecular analysis of genetic variation.

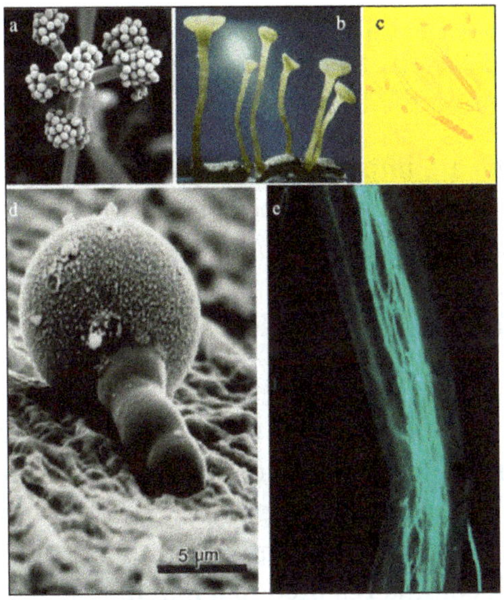


(a) B. cinerea conidiophore with mature conidia in situ (LTSEM). (b) Apothecia of Botryotinia fuckeliana, approximately 10 weeks after spermatization. (c) Two asci each containing eight ascospores, surrounded by ascospores released from damaged asci. (d) Conidium germinating in absence of water droplet on abaxial surface of rose petal (LTSEM reprinted from Williamson et al. Effect of humidity on infection of rose petals by dry-inoculated conidia of Botrytis cinerea. (e) Quiescent infection in raspberry flower: hyphae of B. cinerea growing inside the transmitting tissues in absence of pollen tubes (aniline blue stained and viewed by fluorescence microscopy).

Conidia generated at the sources of primary inoculum follow a well-defined diurnal cycle of initiation, production and dissemination that is regulated by fluctuations in temperature and humidity; a rapid decline in humidity with rise in temperature in early morning causes twisting and drying of conidiophores to eject conidia into air currents, either individually or in small clumps. Water droplets can also disperse conidia, but this is probably not a major dispersal method. Conidia formation is stimulated by specific wavelengths of light and near UV is now generally used to induce sporulation in culture. However, some isolates can sporulate in darkness. Conidia can move on air currents from neighbouring crops, yet most conidia are probably generated from primary sources within the crop. As in many fungi, the conidia contain a self-inhibitor and need to be washed to induce high germination rates in vitro.

B. cinerea shows remarkable flexibility in its use of different environments to germinate and obtain nutrients from a host plant. It was discovered that dry-inoculated conidia produced one or two (sometimes up to five) short germ tubes and no obvious terminal appressoria before effecting entry to an intact host cuticle on leaves, petals or fruits. An extracellular matrix was detected around only the penetration area of the germ tube following dry inoculation of bean leaves and incubation at high humidity, whereas with conidial suspensions much longer germ tubes and extensive secretion of an extracellular matrix was observed. B. cinerea is able to form appressoria, but they are distinct from the classical types found in Colletotrichum or Magnaporthe. B. cinerea germlings do contain melanin in the extracellular matrix which is loosely associated with the fungal cell wall but they do not contain a wall that seals the appressorium from the germ tube, as would be required to enable generating extremely high osmotic pressures. It is therefore not feasible for B. cinerea appressoria to penetrate host tissue by physical pressure alone. If the fungus is growing strongly from a 'saprophytic base' (dead adhering petal, 'bunch trash' in grapes, pollen grains) it can form dome-shaped infection cushions on the host.

In small fruits, the floral organs are important sites for primary infection, but then the pathogen may remain inactive for a considerable period before rapidly destroying tissues at fruit maturity. In grapes there is strong histological evidence that conidia infect mainly the flower receptacle, and to a lesser extent the stigma and styles, and the pathogen is then held in a quiescent state by host defences. Microscopic cracks in the grape cuticle also play a part in later infections, especially if berries are swollen after rain.

The stigmatic fluid in flowers of the 'wet stigma' type serves as a nutrient medium for airborne conidia. For example, in raspberry and strawberry conidia germinate and hyphae grow slowly within the transmitting tissues of styles, following the pathway to the ovules used by pollen tubes where the fungus survives for up to 4 weeks as a saprophyte until the fruit ripens. In the field, dry-inoculation of raspberry flowers with conidia resulted in a halving of the shelf life of apparently healthy fruits developed from them, compared with non-inoculated controls. Petals are important infection sites in many crops, and infected wind-blown petals containing mycelium can be regarded as dispersal propagules in some cases, or important sites of secondary inoculum production, as in Phaseolus vulgaris. Relative humidity (RH) is a crucial environmental factor for B. cinerea, but RH is extremely difficult to regulate experimentally. RH values above 93% are needed for conidial germination and infection of rose petals in the absence of water droplets. Persistence of high RH during blossom periods leads to successive cycles of infection and sporulation, leaving no opportunity for timely removal of ripening fruits and damaging epidemics can result without adequate control measures.

The role of insect vectors for B. cinerea has been recognized only in the last 20 years. In grapes there are several insects known to disperse viable conidia, either on their external appendages or even inside the gut, to deposit inoculum on the surface of fruits. Although B. cinerea is not regarded primarily as a wound pathogen, it can infect wounds and where insects cause damage it can flourish, as in raspberry fruits infested by larvae of raspberry beetle (Byturus tomentosus).

## Molecular Approaches for Analysis of Gene Function

The first successful transformation of a Botrytis strain was achieved by Huang et al. in B. squamosa, but it took several years before the molecular genetics of Botrytis was approached on a broad scale. Today, more than a dozen teams are working intensively on molecular genetics of Botrytis spp. and relevant molecular tools such as transformation protocols, vectors, mutants, and genomic and cDNA libraries are available. B. cinerea has become one of the model systems in molecular phytopathology, also stimulated by the economic damage inflicted and the resulting strong industrial interest. Indeed, the first released genome sequence was determined by an agribusiness company (Syngenta). Due to the growing number of groups interested in basic research on B. cinerea, a non-industrial genome sequencing initiative was started and guided by an international consortium; the sequence will soon be released. Apart from the great perspectives for comparative genomics, transcriptomics, proteomics and evolutionary analyses, the availability of the genome sequence has already significantly stimulated basic research in B. cinerea, because the identification and cloning of genes is no longer a problem. For functional analyses it is also important to estimate the genetic complexity, e.g. how many genes encoding a specific type of enzyme are present?

The high efficiency of targeted gene inactivation allows a rapid functional analysis of putative pathogenicity-related genes. The number of functionally analysed genes is rapidly expanding and even an update of the list presented in Tudzynski and Siewers will soon become obsolete. However, updated information about gene function analysis in B. cinerea will be accessible from a recently established pathogen–host interactions database. In combination with biochemical and cytological methods, these molecular genetic techniques have led to a breakthrough in our understanding of the complex biology of B. cinerea.

## Transformation/Gene Inactivation

Several selection systems are available including resistance cassettes for phleomycin, hygromycin, nourseothricin and glufosinate, allowing the generation of multiple knock-out mutants. Reporter-gene technology is still relatively underdeveloped; there are only a few reports on the successful use of β-glucuronidase or GFP constructs. A feature of B. cinerea transformation is the relatively low efficiency of homologous recombination when using circular vector DNA. Gene inactivation by single cross-over is difficult, whereas gene-replacement approaches using linear transformation cassettes yield high knock-out rates (70–100%). The standard technique of targeted gene inactivation has become gene replacement using long (500–1000 bp) flanks; in most laboratories these constructs are designed by PCR.

For complementation studies or integration of reporter gene constructs requiring ectopic integration, the nitrate reductase system was used successfully. Effective homologous recombination in B. cinerea at the niaD locus was first reported by Levis et al. More recently, complementation and reporter gene constructs were targeted to the niaD locus and homologous integration was monitored by PCR or by chlorate resistance of the transformants.

Good transformation rates can be obtained with the Agrobacterium-T-DNA transfer system, though this system has not yet been used in B. cinerea for targeted gene inactivation, only for insertion mutagenesis. As an alternative for knock-out approaches, RNAi has been used in other phytopathogens for the silencing of genes and it was applied successfully also in B. cinerea. This knock-down approach will be especially valuable for the analysis of gene families and essential genes.

A majority of the knock-out mutants reported so far were derived from strain B05.10. This strain is highly virulent on several host plants and is genetically stable. As it consistently yields high transformation rates, it is now one of the standard recipient strains in most laboratories and it was also used for the first genome sequencing project. Several recent investigations showed that the role of specific genes can differ between B. cinerea strains, e.g. targeted mutation of the pectin methylesterase Bcpme1 reduced virulence in strain Bd90 but not in strain B05. The ability to synthesize the toxin botrydial contributes to virulence in strain T4 but not in B05.10, probably due to the ability of the latter strain to produce a second class of toxins, botcinolides. Mutation of the BOs1 gene encoding a histidine kinase in strain B05.10 leads to an inability to, whereas the same mutant in strain UWS 111 still forms normal appressoria and is able to penetrate. These data emphasize the importance of the choice of strain to be used in such experiments and the urgent need to standardize these parameters in the Botrytis research community. On the other hand, these data show the high degree of flexibility of the pathogen. From this point of view it is interesting to have genome sequences at hand of two strains (T4 and B05.10) that differ considerably in virulence.

## Unbiased Gene Identification

Since the classical 'candidate genes' have more or less been functionally analysed in Botrytis, unbiased cloning approaches gain interest because they offer the perspective of identifying genes with novel functions. Two general strategies are applied. First, screening can be performed for genes that are differentially expressed, e.g. specifically in planta or in certain developmental stages, and these genes can subsequently be analysed for their contribution to virulence by targeted inactivation. Techniques for differential gene expression studies which have been successfully applied in B. cinerea comprise differential screening of cDNA libraries, differential display RT-PCR, suppression subtractive hybridization and macroarrays. In the near future, microarrays will come available and allow whole-genome screening approaches.

A second strategy for identifying virulence genes in an unbiased manner comprises the use of random insertional mutagenesis, which has the distinct advantage of beginning from a known phenotype. While the classical REMI technique that was successfully applied in several phytopathogens did not perform well in B. cinerea, Agrobacterium-mediated transformation has been established for B. cinerea in two laboratories. This system fulfils two major criteria that are essential for use in insertional mutagenesis: most transformants carry single-copy integrations and the integration sites appear to be random. The availability of B. cinerea genomic sequences considerably facilitates the identification of the tagged genes. An insertional mutant library has been established and of the 2800 mutants derived so far, more than 30 are significantly impaired in virulence—a larger collection of virulence mutants than obtained so far with the targeted inactivation approach.

The availability of molecular tools and the option to use '-omics' approaches on a large scale in the near future attracts more research groups, which in turn help to generate new techniques—a self-stimulating process. Thereby, B. cinerea is becoming the most extensively studied necrotrophic pathogen.

## Fusarium Graminearum

The pathogen Fusarium graminearum causes devastating head blight of small grain cereals including wheat and barley. Fusarium head blight (FHB), as a global problem, has great economic impact on the cereal industry due to the reduced grain yield and quality as well as to the contamination by diverse mycotoxins, including deoxynivalenol (DON) and zearalenone, which are harmful for humans and animals.

## Symptoms and Signs

Individual plants of cereal crops (e.g. wheat) produce multiple stems, and each stem produces a single seed spike which emerges at the end of the stem. The spike is composed of multiple spikelets positioned on alternate sides of the spike's stem. Each spikelet is composed of flowering structures where seed develops. The first symptoms of Fusarium head blight occur shortly after flowering. Diseased spikelets exhibit premature bleaching as the pathogen grows and spreads within the head. One or more spikelets located in the top, middle, or bottom of the head may be bleached. Over time, the premature bleaching of the spikelets may progress throughout the entire head.

If the environment is warm and moist, aggregations of light pink/salmon colored spores (sporodochia) may appear on the rachis and glumes of individual spikelets. Later in the season, bluish-black spherical bodies may appear on the surface of affected spikelets. These bodies are sexual structures of the fungus known as perithecia, and can be seen readily in laboratory cultures on carrot agar medium.

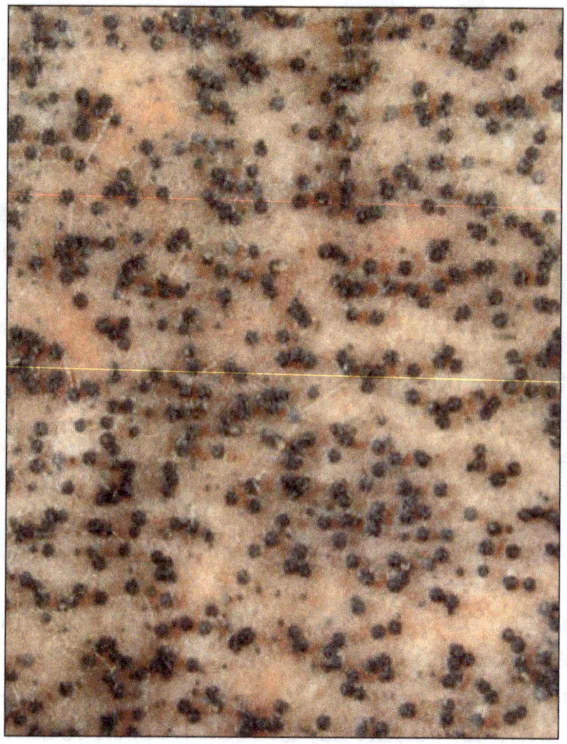

As symptoms progress, the fungus colonizes the developing grain causing it to shrink and wrinkle inside the head. Often, the infected kernels have a rough, shriveled appearance, ranging in color from pink, soft-gray, to light-brown.

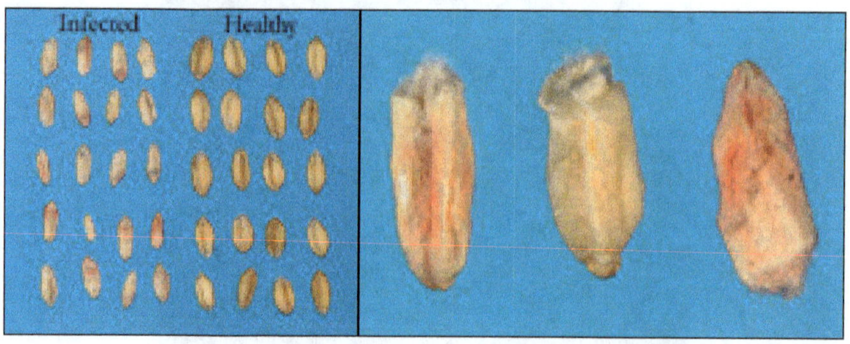


## Pathogen Biology

Fusarium graminearum is an ascomycete, producing sexual spores in a sac known as an ascus (plural asci). The asexual stage of the fungus produces spores called macroconidia, and the sexual stage produces spores called ascospores.

## Asexual Reproduction

The anamorph (asexual stage) of the fungus causing FHB is Fusarium graminearum. Macroconidia (asexual spores) are derived from conidium-producing cells called phialides. The phialides are clustered together in cushion-shaped masses known as sporodochia. The macroconidia are hyaline, canoe-shaped spores usually with five or more septa.

## Sexual Reproduction

The teleomorph (sexual stage) of the fungus is Gibberella zeae. The genus Gibberella belongs in the Hypocreaceae, a family chracterized by brightly colored perithecia that frequently form in stromata (somatic structures on which fruiting bodies develop). The perithecia of G. zeae are a dark blue in color at maturity. Ascospores (sexual spores) form within sacs called asci, and are forcibly discharged from the perithecium through a single small opening known as an ostiole. The ascospores are hyaline to light brown in color, and slightly curved with rounded ends.

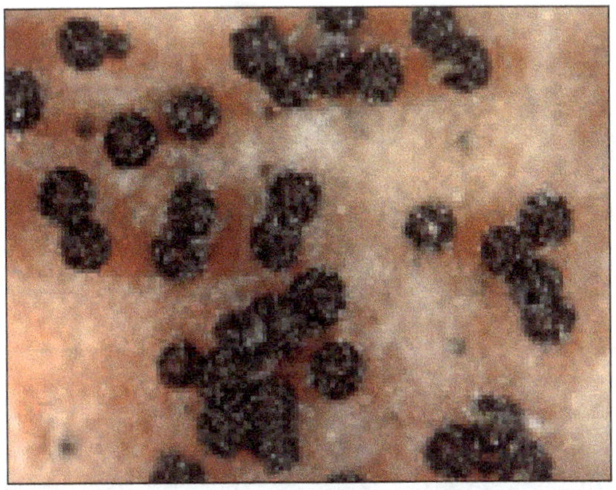

The majority of the isolates of F. graminearum are homothallic, meaning that they are able to sexually reproduce without a partner. Heterothallic isolates, which require a compatible partner for sexual reproduction, are less common. Laboratory studies, however, have demonstrated that many of the homothallic isolates have the ability to outcross with other compatible isolates. The degree to which this occurs in the field under natural conditions is unclear.

## Mycotoxins

Many Fusarium species (including F. graminearum) produce Mycotoxins fungal chemicals that are harmful to animals. These chemicals may operate in nature to disable plant defense mechanisms or to defend the fungus against other microorganisms. The major toxin produced by F. graminearum in association with FHB in wheat and barley is deoxynivalenol (DON). DON is sometimes called vomitoxin because of its deleterious effects on the digestive system of swine and other monogastric animals. DON disrupts normal cell function by inhibiting protein synthesis. Humans consuming flour made from wheat contaminated with DON will often demonstrate symptoms of nausea, fever, headaches, and vomiting.

DON contamination is measured in parts per million (ppm). DON levels in FHB-infected wheat are frequently quite high (>20 ppm). The USDA recommends that DON levels in human foods not exceed 1 ppm. However, individual grain buyers may have lower tolerances of DON in purchased grain. The FDA has various guideline levels of DON permissible in livestock feed: ruminant animals, such as feeder cattle, are the most tolerant, while swine have the highest sensitivity to DON in livestock feed, with pigs refusing feed containing 1ppm of DON.

## Disease Cycle and Epidemiology

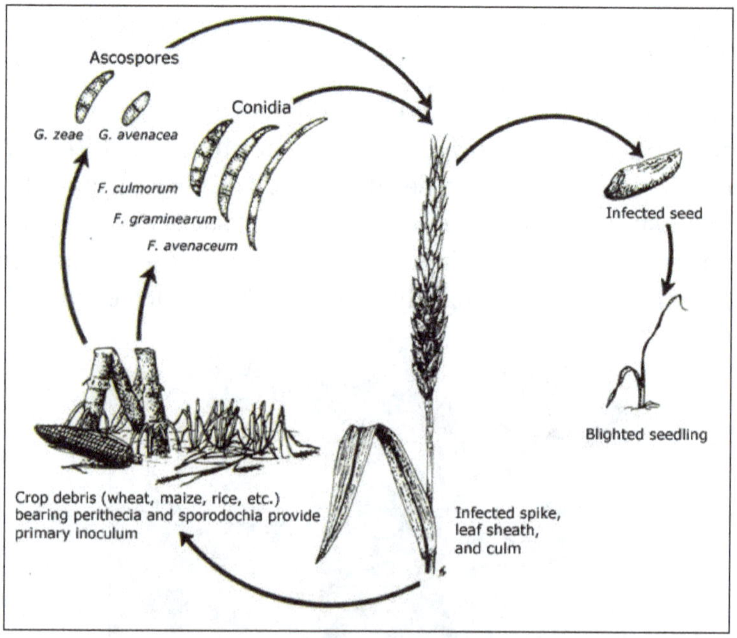


## The Disease Cycle

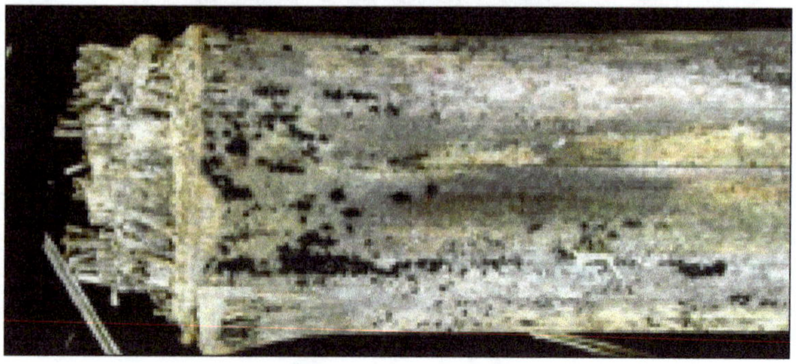

Fusarium graminearum overwinters on infested crop residues (corn stalks, wheat straw, and other host plants). On infested residues, the fungus produces asexual spores (macroconidia) which are dispersed to plants and other plant debris by rain-splash or wind. When conditions are warm, humid, and wet, the sexual stage of the fungus (Gibberella zeae) develops on the infested plant debris. Bluish-black perithecia form on the surface of these residues, and forcibly discharge sexual spores (ascospores) into the air. The ascospores are picked up by turbulent wind currents and may travel great distances in the air.

Infection occurs when the ascospores (and also macroconidia) land on susceptible wheat heads. Extruded anthers during wheat anthesis (flowering) are thought to be the site of primary infection. If the anthers are infected just after their emergence, the fungus will colonize and kill the florets and kernels will not develop. Florets that are infected later will produce diseased kernels that are shriveled and wilted, or "tombstone" in appearance. Kernels that are colonized by the pathogen during late kernel development may not appear to be affected, but may still be contaminated with the mycotoxin DON.

Infected kernels may be used as seed for a subsequent wheat crop. These infected seeds, if left untreated, may give rise to blighted seedlings. The degree to which this occurs in the field depends on the percentage of infected seed and soil conditions affecting the growth and development of the seedlings.

## The Aerobiology of Gibberella Zeae

Viable propagules (ascospores and macroconidia) of G. zeae exist in the air before, during, and after wheat flowering. Generally, the majority of spores dispersed from crop residues travel only short distances, but given appropriate weather conditions and wind, spores may spread over long distances.

Using remote- controlled aircraft and boats equipped with spore-sampling devices researchers are studying the long-range aerial dispersal of the pathogen. Spores of G. zeae have been collected hundreds of feet in the air over agricultural fields, forests, and lakes.

There is considerable debate on the relative importance of local vs. long distance spread of inoculum on disease development. In years with highly favorable weather conditions for pathogen spread and infection, local management of overwintered inoculum (e.g., tillage, spraying of debris, etc.) may have negligible impact on disease levels unless performed over extensive production areas.

## The Environment and FHB

FHB infection is favored by extended periods of high moisture or relative humidity (>90%) and moderately warm temperatures (between 15 to 30 °C/ 59 to 86 °F). These conditions present before, during, and after flowering favor inoculum production, floret infection, and colonization of developing grains.

FHB is well suited for disease forecasting because of its narrow time periods of pathogen sporulation, spore dispersal, and host infection that contribute to epidemic development. Disease forecasting models for FHB have been developed and are widely used. These models incorporate factors such as temperature, humidity, rainfall, inoculum production, and plant development as predictors for the severity of FHB.

## Disease Management of Fusarium Head Blight

### Resistant Cultivars

Since 1990, an extensive research endeavor has focused on development and use of resistant cereal cultivars and integrated pest management systems for the control of Fusarium head blight. Thousands of plant lines are subjected to artificial inoculation with F. graminearum. Those lines having reduced fungal growth and low levels of seed contamination with the mycotoxin DON are selected and advanced in additional breeding trials. To date, sources of resistance conferring complete resistance to FHB have not been identified in wheat. Quantitative Trait Loci (QTL) composed of one or more genes, such as Fhb1 derived from the Chinese wheat cultivar Sumai 3, have been identified in wheat. However, these genes confer only partial resistance to FHB, and many of the initial sources of resistance were not well adapted to most of the grain production regions of the U.S. While some success has been made in transferring FHB resistance from such exotic sources into adapted cultivars, identification and deployment of FHB resistance already present in local native germplasm and cultivars is providing another means to achieve this goal. Ultimately, control of FHB, to meet the very low DON limits in wheat grain, will require an integrated approach including development of cultivars having multiple resistance genes and use of fungicides.

## Agronomic Considerations

Crop sequence (what crops were planted and when) and tillage (soil incorporation of crop residues) have been shown to affect the incidence of FHB. In recent years, decreases in tillage are thought to have contributed to regional scab epidemics by increasing levels of inoculum available for infection.

Since the risk of FHB depends on a viable inoculum source, the management of cereal debris on the soil surface may or may not impact the level of FHB. The relative contribution of inoculum from local and distant sources is not yet fully understood. In regions where there is a significant source of airborne inoculum, local management of the disease (on a single farm) may not be effective.

## Fungicides

Chemical controls, such as fungicides, provide partial control of FHB and associated mycotoxin contamination. A number of foliar fungicides have been used to manage FHB in some areas and are applied around the period of wheat flowering. In many areas, fungicides are rarely used for FHB control because of high cost, variable efficacy, and the erratic nature of FHB epidemics. Research continues to identify fungicides that are more effective for the control of FHB.

Many commercial fungicides that are routinely used for cereal seed treatment also reduce the risk of Fusarium seedling blight. Recently registered fungicides have reduced FHB severity and DON levels in the range of 50-60% across multi-state tests.

## Biological Control

Several investigators are focused on finding affordable and environmentally compatible bio control agents for the management of FHB. Biocontrol agents could play an important role in organic cereal production. In conventional production, such agents may extend protection of spikes past the flowering stage after fungicides can no longer be applied. Certain strains of spore- producing bacteria (such as Bacillus species) and yeasts (such as Cryptococcus flavescens) show some promise for the control of FHB and the reduction of mycotoxin contamination.

### Integrated Management

Integrated management of FHB may one day be achieved by the combined application of bio control agents and fungicides to flowering wheat and barley varieties with partial resistance. Disease forecasting models may help to optimize FHB management by targeting fungicide and bio control applications. The online Fusarium Head Blight Risk Assessment Tool may be used to gauge the relative risk of FHB in wheat fields in the U.S. A public outreach program called Scab Smart provides U.S. wheat and barley growers with the latest information on integrated management tools that can be applied in their section of the country. Disease Forecasting models help producers determine the risk of FHB infection at the flowering period of wheat, and thus help optimize FHB management by having fungicides applied only when models indicate that the risk of FHB infection is present, based on current weather parameters. The spring wheat model also has the option for the producer to choose the level of resistance present in the wheat variety grown. Level of resistance affects the FHB risk.

## Significance

### Losses and Economic Impact

Fusarium head blight is one of the most devastating plant diseases in the world. The United States Department of Agriculture (USDA) ranks FHB as the worst plant disease to hit the US since the rust epidemics in the 1950s. Since 1990, wheat and barley farmers in the United States have lost over $3 billion dollars due to FHB epidemics. Canada has also experienced severe losses since 1990.

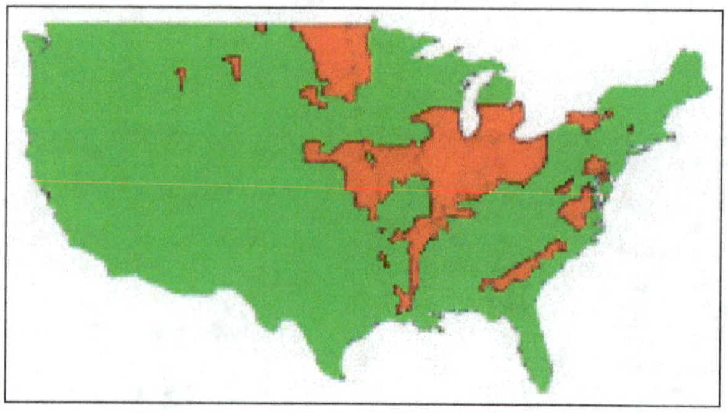

### Mycotoxins as a Threat to Human Health

The mycotoxins produced by F. graminearum may pose a serious threat to human and domestic animal health. Grain that has been infected with the fungus may become incorporated into our staple diets. Strains of the fungus from different countries produce different toxins, some potentially more potent and dangerous than those from strains currently in the United States.

### The Years Ahead

Many fear that the disease will continue to be the cause of major crop losses in the United States and beyond, with the potential to put a major dent in the world's food supply during the years to come. Currently, low commodity prices, increasingly stringent tolerances for mycotoxins, and lack of affordable and highly efficacious control methods, have made wheat and barley production unprofitable in many areas of North America. Who will grow the grain to produce our bread, pastries, pasta, and beer in the years ahead?

## References


- Food-supply, teaching-materials, integrate-1230: serc.carleton.edu, Retrieved 05, May 2020
- Plant-pathogens, plant-growth: intechopen.com, Retrieved 17, March 2020
- Bacteria, prokaryote, disandpath, edcenter: apsnet.org, Retrieved 14, August 2020
- BacterialPlantPathogens-001: s3.wp.wsu.edu, Retrieved 13, April 2020
- Pseudomonas-syringae, agricultural-and-biological-sciences: sciencedirect.com, Retrieved 09, July 2020
- Xanthomonas-species-(Bacterial-leaf-spot-on-ornamentals): wiki.bugwood.org, Retrieved 11, January 2020
- PlantViruses, disandpath, edcenter: apsnet.org, Retrieved 04, March 2020
- Tobacco-mosaic-virus: extension.psu.edu, Retrieved 22, June 2020
- Cucumbermosaic, disandpath, edcenter: apsnet.org, Retrieved 18, February 2020

# Chapter 3

## Common Diseases in Plants

A plant disease refers to any disorder that impairs the normal functioning of a plant or its related systems. Plant diseases are usually infectious and may affect a large plant population if not managed properly. This may result in massive commercial and population losses. This chapter is a compendium of some important plant diseases commonly found in plants.

Plant disease is an impairment of the normal state of a plant that interrupts or modifies its vital functions. All species of plants, wild and cultivated alike, are subject to disease. Although each species is susceptible to characteristic diseases, these are, in each case, relatively few in number. The occurrence and prevalence of plant diseases vary from season to season, depending on the presence of the pathogen, environmental conditions, and the crops and varieties grown. Some plant varieties are particularly subject to outbreaks of diseases while others are more resistant to them.

### Nature and Importance of Plant Diseases

Loss of crops from plant diseases may also result in hunger and starvation, especially in less-developed countries where access to disease-control methods is limited and annual losses of 30 to 50 percent are not uncommon for major crops. In some years, losses are much greater, producing catastrophic results for those who depend on the crop for food. Major disease outbreaks among food crops have led to famines and mass migrations throughout history. The devastating outbreak of late blight of potato (caused by the water mold Phytophthora infestans) that began in Europe in 1845 brought about the Great Famine that caused starvation, death, and mass migration of the Irish. Of Ireland's population of more than eight million, approximately one million (about 12.5 percent) died of starvation or famine-related illness, and 1.5 million (almost 19 percent) emigrated, mostly to the United States, as refugees from the destructive blight. This water mold thus had a tremendous influence on economic, political, and cultural development in Europe and the United States. During World War I, late blight damage to the potato crop in Germany may have helped end the war.

### Diseases—A Normal Part of Nature

Plant diseases are a normal part of nature and one of many ecological factors that help keep the hundreds of thousands of living plants and animals in balance with one another. Plant cells contain special signaling pathways that enhance their defenses against insects, animals, and pathogens. One such example involves a plant hormone called jasmonate (jasmonic acid). In the absence of harmful stimuli, jasmonate binds to special proteins, called JAZ proteins, to regulate plant growth, pollen production, and other processes. In the presence of harmful stimuli, however, jasmonate switches its signaling pathways, shifting instead to directing processes involved in boosting plant defense. Genes that produce jasmonate and JAZ proteins represent potential targets for genetic engineering to produce plant varieties with increased resistance to disease.

Humans have carefully selected and cultivated plants for food, medicine, clothing, shelter, fibre, and beauty for thousands of years. Disease is just one of many hazards that must be considered when plants are taken out of their natural environment and grown in pure stands under what are often abnormal conditions.

Many valuable crop and ornamental plants are very susceptible to disease and would have difficulty surviving in nature without human intervention. Cultivated plants are often more susceptible to disease than are their wild relatives. This is because large numbers of the same species or variety, having a uniform genetic background, are grown close together, sometimes over many thousands of square kilometres. A pathogen may spread rapidly under these conditions.

## Definitions of Plant Disease

In general, a plant becomes diseased when it is continuously disturbed by some causal agent that results in an abnormal physiological process that disrupts the plant's normal structure, growth, function, or other activities. This interference with one or more of a plant's essential physiological or biochemical systems elicits characteristic pathological conditions or symptoms.

Plant diseases can be broadly classified according to the nature of their primary causal agent, either infectious or noninfectious. Infectious plant diseases are caused by a pathogenic organism such as a fungus, bacterium, mycoplasma, virus, viroid, nematode, or parasitic flowering plant. An infectious agent is capable of reproducing within or on its host and spreading from one susceptible host to another. Noninfectious plant diseases are caused by unfavorable growing conditions, including extremes of temperature, disadvantageous relationships between moisture and oxygen, toxic substances in the soil or atmosphere, and an excess or deficiency of an essential mineral. Because noninfectious causal agents are not organisms capable of reproducing within a host, they are not transmissible.

In nature, plants may be affected by more than one disease-causing agent at a time. A plant that must contend with a nutrient deficiency or an imbalance between soil moisture and oxygen is often more susceptible to infection by a pathogen, and a plant infected by one pathogen is often prone to invasion by secondary pathogens. The combinations of all disease-causing agents that affect a plant make up the disease complex. Knowledge of normal growth habits, varietal characteristics, and normal variability of plants within a species—as these relate to the conditions under which the plants are growing is required for a disease to be recognized.

The study of plant diseases is called plant pathology. Pathology is derived from the two Greek words pathos (suffering, disease) and logos (discourse, study). Plant pathology thus means a study of plant diseases.

## Disease Development and Transmission

### Pathogenesis and Saprogenesis

Pathogenesis is the stage of disease in which the pathogen is in intimate association with living host tissue. Three fairly distinct stages are involved:

1. Inoculation: Transfer of the pathogen to the infection court, or area in which invasion of the plant occurs (the infection court may be the unbroken plant surface, a variety of wounds,

or natural openings—e.g., stomata microscopic pores in leaf surfaces, hydathodes stoma-ta-like openings that secrete water, or lenticels small openings in tree bark).

2. Incubation: The period of time between the arrival of the pathogen in the infection court and the appearance of symptoms.
3. Infection: The appearance of disease symptoms accompanied by the establishment and spread of the pathogen.

One of the important characteristics of pathogenic organisms, in terms of their ability to infect, is virulence. Many different properties of a pathogen contribute to its ability to spread through and to destroy the tissue. Among these virulence factors are toxins that kill cells, enzymes that destroy cell walls, extracellular polysaccharides that block the passage of fluid through the plant system, and substances that interfere with normal cell growth. Not all pathogenic species are equal in virulence—that is, they do not produce the same amounts of the substances that contribute to the invasion and destruction of plant tissue. Also, not all virulence factors are operative in a particular disease. For example, toxins that kill cells are important in necrotic diseases, and enzymes that destroy cell walls play a significant role in soft rot diseases.

Many pathogens, especially among the bacteria and fungi, spend part of their life cycle as pathogens and the remainder as saprotrophs.

Saprogenesis is the part of the pathogen's life cycle when it is not in vital association with living host tissue and either continues to grow in dead host tissue or becomes dormant. During this stage, some fungi produce their sexual fruiting bodies; the apple scab (Venturia inaequalis), for example, produces perithecia, flask-shaped spore-producing structures, in fallen apple leaves. Other fungi produce compact resting bodies, such as the sclerotia formed by certain root- and stem-rotting fungi (Rhizoctonia solani and Sclerotinia sclerotiorum) or the ergot fungus (Claviceps purpurea). This resting body, which are resistant to extremes in temperature and moisture, enable the pathogen to survive for months or years in soil and plant debris in the absence of a living host.

## Epiphytotics

When the number of individuals a disease affects increases dramatically, it is said to have become epidemic (meaning "on or among people"). A more precise term when speaking of plants, however, is epiphytotic ("on plants"); for animals, the corresponding term is epizootic. In contrast, endemic (enphytotic) diseases occur at relatively constant levels in the same area each year and generally cause little concern.

Epiphytotics affect a high percentage of the host plant population, sometimes across a wide area. They may be mild or destructive and local or regional in occurrence. Epiphytotics result from various combinations of factors, including the right combination of climatic conditions. An epiphytotic may occur when a pathogen is introduced into an area in which it had not previously existed. Examples of this condition include the downy mildews (Sclerospora species) and rusts (Puccinia species) of corn in Africa during the 1950s, the introduction of the coffee rust fungus into Brazil in the 1960s, and the entrance of chestnut blight (Endothia parasitica) into the United States shortly after 1900. Also, when new plant varieties are produced by plant breeders without regard for all enphytotic diseases that occur in the same area to some extent each year (but which are normally of minor importance), some of these varieties may prove very susceptible to previously unimportant pathogens. Examples of this situation include the development of oat varieties with Victoria parentage, which, although highly resistant to rusts and smuts, proved very susceptible to Helminthosporium blight, formerly a minor disease of grasses. The destructiveness of this disease resulted in a major shift of oat varieties in the United States in the mid-1940s. Corn (maize) with male-sterile cytoplasm (i.e., plants with tassels that do not extrude anthers or pollen), grown on 60 million acres (24 million hectares) in the United States, was attacked in 1970 by a virulent new race of the southern corn leaf blight fungus, resulting in a loss of about 700 million bushels of corn. More recently the new Helminthosporium race was widely disseminated and was reported from most continents. Finally, epiphytotics may occur when host plants are cultivated in large acreages where previously little or no land was devoted to that crop.

Epiphytotics may occur in cycles. When a plant disease first appears in a new area, it may grow rapidly to epiphytotic proportions. In time, the disease wanes, and, unless the host species has been completely wiped out, the disease subsides to a low level of incidence and becomes enphytotic. This balance may change dramatically by conditions that favor a renewed epiphytotic. Among such conditions are weather (primarily temperature and moisture), which may be very favorable for multiplication, spread, and infection by the pathogen; introduction of a new and more susceptible host; development of a very aggressive race of the pathogen; and changes in cultural practices that create a more favorable environment for the pathogen.

## Environmental Factors Affecting Disease Development

Important environmental factors that may affect development of plant diseases and determine whether they become epiphytotic include temperature, relative humidity, soil moisture, soil pH, soil type, and soil fertility.

### Plant Disease: Infection from Water

Learn how rainwater droplets, splashing from leaf to leaf, infect plants with disease.

### Temperature

Each pathogen has an optimum temperature for growth. In addition, different growth stages of fungi, such as the production of spores (reproductive units), their germination, and the growth of the mycelium (the filamentous main fungus body), may have slightly different optimum temperatures.

Storage temperatures for certain fruits, vegetables, and nursery stock are manipulated to control fungi and bacteria that cause storage decay, provided the temperature does not change the quality of the products. Little, except limited frost protection, can be done to control air temperature in fields, but greenhouse temperatures can be regulated to check disease development.

Knowledge of optimum temperatures, usually combined with optimum moisture conditions, permits forecasting, with a high degree of accuracy, the development of such diseases as blue mold of tobacco (Peronospora tabacina), downy mildews of vine crops (Pseudoperonospora cubensis) and lima beans (Phytophthora phaseoli), late blight of potato and tomato (Phytophthora infestans), leaf spot of sugar beets (Cercospora beticola), and leaf rust of wheat (Puccinia recondita tritici). Effects of temperature may mask symptoms of certain viral and mycoplasmal diseases, however, making them more difficult to detect.

## Relative Humidity

Relative humidity is very critical in fungal spore germination and the development of storage rots. Rhizopus soft rot of sweet potato (Rhizopus stolonifer) is an example of a storage disease that does not develop if relative humidity is maintained at 85 to 90 percent, even if the storage temperature is optimum for growth of the pathogen. Under these conditions, the sweet potato root produces suberized (corky) tissues that wall off the Rhizopus fungus.

High humidity favours development of the great majority of leaf and fruit diseases caused by fungi, water molds, and bacteria. Moisture is generally needed for spore germination, the multiplication and penetration of bacteria, and the initiation of infection. Germination of powdery mildew spores occurs best at 90 to 95 percent relative humidity. Diseases in greenhouse crops—such as leaf mold of tomato (Cladosporium fulvum) and decay of flowers, leaves, stems, and seedlings of flowering plants, caused by Botrytis species—are controlled by lowering air humidity or by avoiding spraying plants with water.

## Soil Moisture

High or low soil moisture may be a limiting factor in the development of certain root rot diseases. High soil-moisture levels favor development of destructive water mold fungi, such as species of Aphanomyces, Pythium, and Phytophthora. Excessive watering of houseplants is a common problem. Overwatering, by decreasing oxygen and raising carbon dioxide levels in the soil, makes roots more susceptible to root-rotting organisms.

## Potato Scab

Common scab of potato:

Diseases such as take-all of cereals (Ophiobolus graminis); charcoal rot of corn, sorghum, and soybean (Macrophomina phaseoli); common scab of potato (Streptomyces scabies); and onion white rot (Sclerotium cepivorum) are most severe under low soil-moisture levels.

## Soil pH

Soil pH, a measure of acidity or alkalinity, markedly influences a few diseases, such as common scab of potato and clubroot of crucifers (Plasmodiophora brassicae). Growth of the potato scab organism is suppressed at a pH of 5.2 or slightly below (pH 7 is neutral; numbers below 7 indicate acidity, and those above 7 indicate alkalinity). Scab is not normally a problem when the natural soil pH is about 5.2. Some farmers add sulfur to their potato soil to keep the pH about 5.0. Clubroot of crucifers (members of the mustard family, including cabbage, cauliflower, and turnips), on the other hand, can usually be controlled by thoroughly mixing lime into the soil until the pH becomes 7.2 or higher.

## Soil Type

Certain pathogens are favoured by loam soils and others by clay soils. Phymatotrichum root rot attacks cotton and some 2,000 other plants in the southwestern United States. This fungus is serious only in black alkaline soils—pH 7.3 or above—that are low in organic matter. Fusarium wilt disease, which attacks a wide range of cultivated plants, causes more damage in lighter and higher (topographically) soils. Nematodes are also most damaging in lighter soils that warm up quickly.

## Soil Fertility

Greenhouse and field experiments have shown that raising or lowering the levels of certain nutrient elements required by plants frequently influences the development of some infectious diseases—for example, fire blight of apple and pear, stalk rots of corn and sorghum, Botrytis blights, Septoria diseases, powdery mildew of wheat, and northern leaf blight of corn. These diseases and many others are more destructive after application of excessive amounts of nitrogen fertilizer. This condition can often be counteracted by adding adequate amounts of potash, a fertilizer containing potassium.

## Requirements for Disease Development

Infectious disease cannot develop if any one of the following three basic conditions is lacking: (1) the proper environment, the most important environmental factors being the amount and frequency of rains or heavy dews, the relative humidity, and the air and soil temperatures, (2) the presence of a virulent pathogen, and (3) a susceptible host. Effective disease-control measures are aimed at breaking this environment-pathogen-host triangle. Loss resulting from disease is reduced, for example, if the host can be made more resistant or immune through such techniques as plant breeding or genetic engineering. In addition, the environment can be made less favorable for invasion by the pathogen and more favorable for the growth of the host plant. Finally, the pathogen can be killed or prevented from reaching the host. These basic methods of control can be divided into a number of cultural, chemical, and biological practices to help control the disease.

## Diagnosis of Plant Diseases

Rapid and accurate diagnosis of disease is necessary before proper control measures can be suggested. It is the first step in the study of any disease. Diagnosis is largely based on characteristic symptoms expressed by the diseased plant. Identification of the pathogen is also essential to diagnosis.

Three steps involved in diagnosis include careful observation and classification of the facts, evaluation of the facts, and a logical decision as to the cause.

## Variable Factors Affecting Diagnosis

A skilled diagnostician must know the normal appearance of an affected plant species, its local air and soil environment, the cultural conditions under which it is growing, the pathogens described for the area, and the disease-developing potential of the pathogen. Diagnosis is best done in the presence of the growing plant. Disease is suspected when, for example, part or all of a plant begins to die. Disease also is indicated when blossoms, leaves, stems, roots, or other plant parts appear abnormal—i.e., misshapen, curled, discolored, overdeveloped, or underdeveloped. Diseased plants also often fail to respond normally to fertilizing, watering, pruning, insect and mite control, or other recommended practices.

Conditions other than infection with a pathogen, however, may produce similar or identical symptoms. Some of these have been described, but numerous other conditions must be considered as well when plants are adversely affected. For example, an affected plant may not be adapted to the area in which it is growing. It may not be able to withstand the extremes in soil moisture, temperature, wind, light, or humidity of the local situation. Damage to plants may be caused by insects, mites, rodents, pets, or humans. The soil may be poorly drained, gravelly, or overly sandy; it may be covering buried debris such as boards, cement blocks, bricks, or mortar; or it may be too dry or otherwise unfavorable for good plant growth. Problems also are caused by high winds, hail, lightning, blowing sand, a heavy load of snow or ice, flooding, fire, ice-removal chemicals, mechanical injury by garden tools or machinery, and fumes from weed-killing chemicals, trash burners, nearby industrial plants, or motor vehicles. The affected plant may have received treatment different from nearby healthy ones—watering, fertilizing, pest control, pruning, or depth of planting, for example. If different species or kinds of plants in the same area have similar symptoms, the chances are that a pathogen is not involved. Most infectious diseases are highly specific for individual or closely related plant species, and similar symptoms on unrelated plants are usually an indication of some environmental factor rather than a disease-causing organism.

Examination of leaves is usually considered to be the best starting point in diagnosis. The color, size, shape, and margins of spots and blights (lesions) are often associated with a particular fungus or bacterium. Many fungi produce "signs" of disease, such as mold growth or fruiting bodies that appear as dark specks in the dead area. Early stages of bacterial infections that develop on leaves or fruits during humid weather often appear as dark and water-soaked spots with a distinct margin and sometimes a halo—a lighter-colored ring around the spot.

Low winter temperatures and late spring or early fall freezes cause blasting (sudden death) of leaf and flower buds or sudden blighting (discoloration and death) of tender foliage.

Insect-injured leaves usually show evidence of feeding, such as holes, discoloration, stippling, blotching, downward curling, or other deformations.

Scorching of leaf margins and between the veins is common following hot, dry, windy weather. Similar symptoms are produced by an excess of water, an imbalance of essential nutrients, an excess of soluble salts, changes in the soil water table or soil grade, gas or fume injury, and root injury or disease.

Viral diseases, such as mosaics and yellows, are sometimes confused with injury by a hormone-type weed killer, unbalanced nutrition, and soil that is excessively alkaline or acid. Nearby plant species are often examined to see if similar symptoms are evident on several different types of plants.

Examination of stems, shoots, branches, and trunk follows a thorough leaf examination. Sunken, swollen, or discolored areas in the fleshy stem or bark may indicate canker infection by a fungus or bacterium or injury caused by excessively high or low temperatures, hail, tools, equipment, vehicles, or girdling wires.

Fruiting bodies of fungi in or on such areas often indicate secondary infection. Accurate identification of signs as belonging to a pathogenic organism or a secondary or saprophytic one is difficult. Tissues directly infected by pathogenic fungi or bacteria normally show a gradual change in color or consistency. Injuries, in comparison, are usually well defined with an abrupt change from healthy to affected tissue.

Holes and sawdustlike debris are evidence of boring insects that usually invade woody plants in a low state of vigour. Other borer indications include wilting and dieback (progressive death of shoots that begins at tip and works downward). These symptoms also are produced by fungi and bacteria that invade water- and food-conducting vascular tissue.

Symptoms of wilt-inducing microorganisms include dark streaks in sapwood of wilted branches when the wood is cut through at an angle.

Abnormal suckers or water sprouts on trees can indicate careless pruning, extremes in temperature or water supply, structural injury, or disease.

Galls, which are unsightly overgrowths on stem, branch, or trunk, may indicate crown gall, insect injury, water imbalance between plant and soil, or other factors. Crown gall is infectious and develops as rough, roundish galls at wounds, resulting from grafting, pruning, or cultivating.

## Crown Gall

Plant infected with crown gall:

Wood-decay fungi also enter unprotected wounds, resulting in discolored, water-soaked, spongy, stringy, crumbly, or hard rots of living and dead wood. External evidence of wood-decay fungi are clusters of mushrooms (or toadstools) and hoof- or shelf-shaped fungal fruiting structures, called conks, punks, or brackets.

Aboveground symptoms of many root problems look alike. They include stunting of leaf and twig growth, poor foliage color, gradual or sudden decline in vigor and productivity, shoot wilting and dieback, and even rapid death of the plant. The causes include infectious root and crown rot; nematode, insect, or rodent feeding; low temperature or lightning injury; household gas injury; poor soil type or drainage; change in soil grade; or massive removal of roots in digging utility trenches and construction.

Abnormal root growth is revealed by comparison with healthy roots. Some nematodes, such as root knot (Meloidogyne species), produce small to large galls in roots; other species cause affected roots to become discolored, stubby, excessively branched, and decayed. Bacterial and fungal root rots commonly follow feeding by nematodes, insects, and rodents.

Diagnosis of a disease complex, one with two or more causes, is usually difficult and requires separation and identification of the individual causes.

## Symptoms

The variety of symptoms, the internal and external expressions of disease, that result from any disease form the symptom complex, which, together with the accompanying signs, makes up the syndrome of the disease.

Generalized symptoms may be classified as local or systemic, primary or secondary, and microscopic or macroscopic. Local symptoms are physiological or structural changes within a limited area of host tissue, such as leaf spots, galls, and cankers. Systemic symptoms are those involving the reaction of a greater part or all of the plant, such as wilting, yellowing, and dwarfing. Primary symptoms are the direct result of pathogen activity on invaded tissues (e.g., swollen "clubs" in clubroot of cabbage and "galls" formed by feeding of the root knot nematode). Secondary symptoms result from the physiological effects of disease on distant tissues and uninvaded organs (e.g., wilting and drooping of cabbage leaves in hot weather resulting from clubroot or root knot). Microscopic disease symptoms are expressions of disease in cell structure or cell arrangement seen under a microscope. Macroscopic symptoms are expressions of disease that can be seen with the unaided eye. Specific macroscopic symptoms are classified under one of four major categories: prenecrotic, necrotic, hypoplastic, and hyperplastic or hypertrophic. These categories reflect abnormal effects on host cells, tissues, and organs that can be seen without a hand lens or microscope.

## Clubroot

Cauliflower roots infected with clubroot.

## Signs

Besides symptoms, the diagnostician recognizes signs characteristic of specific diseases. Signs are either structures formed by the pathogen or the result of interaction between pathogen and host—e.g., ooze of fire blight bacteria, slime flux from wetwood of elm, odor of tissues affected with bacterial soft rot.

## Technological Advances in the Identification of Pathogenic Agents

Developments in microscopy, serology and immunology, molecular biology, and laboratory instrumentation have resulted in many new and sophisticated laboratory procedures for the identification of plant pathogens, particularly bacteria, viruses, and viroids. The techniques of traditional scanning microscopy and transmission electron microscopy have been applied to immunosorbent electron microscopy, in which the specimen is subject to an antigen-antibody reaction before observation and scanning tunneling microscopy, which provides information about the surface of a specimen by constructing a three-dimensional image.

Serological tests have been made more specific and convenient to perform since the discovery of a technique to produce large quantities of monoclonal antibodies, which bind to only one specific antigen. The sensitivity of antigen-antibody detection has been significantly increased by a radioimmunoassay (RIA) procedure. In this procedure a "known" antigen is overlayed on a plastic plate to which antigen molecules adhere. A solution of antibody is applied to the same plate; if the antibody is specific to the antigen, it will combine with it. This is followed by the application of radioactively labeled anti-antibody, which is allowed to react and then washed off. The radioactivity that remains on the plate is a measure of the amount of antibody that combined with the known fixed antigen. Another highly sensitive immunoassay is the enzyme-linked immunosorbent assay (ELISA). In principle this assay is similar to the RIA except that an enzyme system, instead of radioactivity, is used as an indicator of an antigen-antibody combination.

New analytic methods in molecular biology have made genetic studies for the characterization and identification of bacteria more practical. The DNA hybridization technique is an example. A strand of DNA from a known species (the probe) is radioactively labeled and "mixed" with DNA from an unidentified species. If the probe and the unknown DNA are from identical species, they will have complementary DNA sequences that enable them to bind to one another. Bound to DNA from the unknown species, the probe acts as a marker and identifies the bacteria.

The growing demand for quick identification of microorganisms has resulted in the development of instrumentation for automated technology that allows a large number of tests to be performed on many specimens in a short period of time. The results are read automatically and analyzed by a computer program to identify the pathogens.

## Principles of Disease Control

Successful disease control requires thorough knowledge of the causal agent and the disease cycle, host-pathogen interactions in relation to environmental factors, and cost. Disease control starts with the best variety, seed, or planting stock available and continues throughout the life of the plant. For harvested crops, disease control extends through transport, storage, and marketing. Relatively few diseases are controlled by a single method; the majority require several approaches. These often need to be integrated into a broad program of biological, cultural, and chemical methods to control as many different pests—including insects, mites, rodents, and weeds—on a given crop as possible.

Most control measures are directed against inoculum of the pathogen and involve the principles of exclusion and avoidance, eradication, protection, host resistance and selection, and therapy.

## Exclusion and Avoidance

The principle of exclusion and avoidance is to keep the pathogen away from the growing host plant. This practice commonly excludes pathogens by disinfection of plants, seeds, or other parts, using chemicals or heat. Inspection and certification of seed and other planting stock help ensure freedom from disease. For gardeners this involves sorting bulbs or corms before planting and rejecting diseased plants. Federal and state plant quarantines, or embargoes, have been established to prevent introduction of potentially destructive pathogens into areas currently free of the disease. More than 150 countries now have established quarantine regulations.

## Eradication

Eradication is concerned with elimination of the disease agent after it has become established in the area of the growing host or has penetrated the host. Such measures include crop rotation, destruction of the diseased plants, and elimination of alternate host plants, pruning, disinfection, and heat treatments.

Crop rotation with nonsusceptible crops "starves out" bacteria, fungi, and nematodes with a restricted host range. Some pathogens can survive only as long as the host residue persists, usually no more than a year or two. Many pathogens, however, are relatively unaffected by rotation because they become established as saprotrophs in the soil (e.g., Fusarium and Pythium species;

Rhizoctonia solani; and the potato scab actinomycete, Streptomyces scabies) or their propagative structures remain dormant but viable for many years (e.g., cysts of cyst nematodes, sporangia of the cabbage clubroot fungus, and onion smut spores).

Burning, deep plowing of plant debris, and fall spraying are used against such diseases as leaf blights of tomato, Dutch elm disease, and apple scab. Destruction of weed hosts also helps control such viral diseases as cucumber mosaic and curly top. For fungi whose complete life cycle requires two different host species, such as black stem rust of cereals and white-pine blister rust, destruction of alternate hosts is effective. Destruction of diseased plants helps control Dutch elm disease, oak wilt, and peach viral diseases—mosaic, phony peach, and rosette. Elimination of citrus canker in the southeastern United States has been one of the few successful eradication programs in history. Infected trees were sprayed with oil and burned.

Pruning and excision of a diseased portion of the plant have aided in reducing inoculum sources for canker and wood-rot diseases of shade trees and fire blight of pome fruits. Disinfection of contaminated tools, as well as packing and shipping containers, controls a wide range of diseases. Direct application of dry or wet heat is used to obtain seeds, bulbs, other propagative materials, and even entire plants free of viruses, nematodes, and other pathogens.

## Protection

The principle of protection involves placing a barrier between the pathogen and the susceptible part of the host to shield the host from the pathogen. This can be accomplished by regulation of the environment, cultural and handling practices, control of insect carriers, and application of chemical pesticides.

## Regulation of the Environment

Selection of outdoor growing areas where weather is unfavorable for disease is a method of controlling disease by regulating the environment. Control of viral diseases of potato, for example, can be accomplished by growing the seed crop in northern regions where low temperatures are unfavorable for the aphid carriers. Another environmental factor that can be brought under control is the storage and in-transit environment. A variety of postharvest diseases of potato, sweet potato, onion, cabbage, apple, pear, and other crops are controlled in storage and shipment by keeping humidity and temperature low and by reducing the quantity of ethylene and other natural gases in storage houses.

## Cultural Practices

Selection of the best time and depth of seeding and planting is an effective cultural practice that reduces disease impact. Shallow planting of potatoes may help to prevent Rhizoctonia canker. Early fall seeding of winter wheat may be unfavorable for seedling infection by wheat bunt teliospores. Cool-temperature crops can be grown in soils infested with root knot nematode and harvested before soil temperatures become favorable for nematode activity. Adjustment of soil moisture is another cultural practice of widespread usefulness. For example, seed decay, damping-off (the destruction of seedlings at the soil line), and other seedling diseases are favoured by excessively wet soils. The presence of drain tiles in poorly drained fields and the use of ridges or beds for plants are often beneficial. Adjustment of soil pH also leads to control of some diseases. Common potato scab can be controlled by adjusting the pH to 5.2 or below; other acid-tolerant plants then must be used in crop rotation, however.

## Regulation of Fertility Level and Nutrient Balance

Potash and nitrogen, and the balance between the two, may affect the incidence of certain bacterial, fungal, and viral diseases of corn, cotton, tobacco, and sugar beet. A number of microelements, including boron, iron, zinc, manganese, magnesium, copper, sulfur, and molybdenum, may cause noninfectious diseases of many crop and ornamental plants. Adjusting the soil pH, adding chelated (bound or enclosed in large organic molecules) or soluble salts to the soil, or spraying the foliage with these or similar salts is a corrective measure.

## Handling Practices

Late blight on potato tubers can be controlled by delaying harvest until the foliage has been killed by frost, chemicals, or mechanical beaters. Avoidance of bruises and cuts while digging, grading, and packing potatoes, sweet potatoes, and bulb crops also reduces disease incidence.

## Control of Insect Vectors

There are many examples in which losses by bacteria, viruses, and mycoplasma-like disease agents can be reduced by controlling aphids, leafhoppers, thrips, beetles, and other carriers of these agents. Insect vectors can be controlled with organic or synthetic insecticides or by means of biological control.

## Chemical Control

A variety of chemicals are available that have been designed to control plant diseases by inhibiting the growth of or by killing the disease-causing pathogens. Chemicals used to control bacteria (bactericides), fungi (fungicides), and nematodes (nematicides) may be applied to seeds, foliage, flowers, fruit, or soil. They prevent or reduce infections by utilizing various principles of disease control. Eradicants are designed to kill a pathogen that may be present in the soil, on the seeds, or on vegetative propagative organs, such as bulbs, corms, and tubers. Protectants place a chemical barrier between the plant and the pathogen. Therapeutic chemicals are applied to combat an infection in progress.

Soil treatments are designed to kill soil-inhabiting nematodes, fungi, and bacteria. This eradication can be accomplished using steam or chemical fumigants. Soilborne nematodes can be killed by applying granular or liquid nematicides. Most soil is treated well before planting; however, certain fungicides can be mixed with the soil at planting time.

Seeds, bulbs, corms, and tubers are frequently treated with chemicals to eradicate pathogenic bacteria, fungi, and nematodes and to protect the seeds against organisms in the soil—mainly fungi—that cause decay and damping-off. Seeds are often treated with systemic fungicides, which are absorbed and provide protection for the growing seedling.

Protective sprays and dusts applied to the foliage and fruit of crops and ornamentals include a wide range of organic chemicals designed to prevent infection. Protectants are not absorbed by or translocated through the plant; thus they protect only those parts of the plant treated before invasion by the pathogen. A second application is often necessary because the chemical may be removed by wind, rain, or irrigation or may be broken down by sunlight. New, untreated growth also is susceptible to infection. New chemicals are constantly being developed.

## Biological Control

Biological control of plant diseases involves the use of organisms other than humans to reduce or prevent infection by a pathogen. These organisms are called antagonists; they may occur naturally within the host's environment, or they may be purposefully applied to those parts of the potential host plant where they can act directly or indirectly on the pathogen.

Although the effects of biological control have long been observed, the mechanisms by which antagonists achieve control is not completely understood. Several methods have been observed: some antagonists produce antibiotics that kill or reduce the number of closely related pathogens; some are parasites on pathogens; and others simply compete with pathogens for available food.

Cultural practices that favor a naturally occurring antagonist and exploit its beneficial action often are effective in reducing disease. One technique is to incorporate green manure, such as alfalfa, into the soil. Saprotrophic microorganisms feed on the green manure, depriving potential pathogens of available nitrogen. Another practice is to make use of suppressive soils—those in which a pathogen is known to persist but causes little damage to the crop. A likely explanation for this phenomenon is that suppressive soils harbour antagonists that compete with the pathogen for food and thereby limit the growth of the pathogen population.

Other antagonists produce substances that inhibit or kill potential pathogens occurring in close proximity. An example of this process, called antibiosis, is provided by marigold (Tagetes species) roots, which release terthienyls, chemicals that are toxic to several species of nematodes and fungi.

Only a few antagonists have been developed specifically for use in plant-disease control. Citrus trees are inoculated with an attenuated strain of tristeza virus, which effectively controls the virulent strain that causes the disease. An avirulent strain of Agrobacterium radiobacter (K84) can be applied to plant wounds to prevent crown gall caused by infection with Agrobacterium tumefaciens. Many more specific antagonists are being investigated and hold much promise for future control of disease.

## Therapy

Therapeutic measures have been used much less often in plant pathology than in human or animal medicine. The recent development of systemic fungicides such as oxathiins, benzimidazoles, and pyrimidines has enabled growers to treat many plants after an infection has begun. Systemic chemicals are absorbed by and translocated within the plant, restricting the spread and development of pathogens by direct or indirect toxic effects or by increasing the ability of the host to resist infection.

Antibiotics have been developed to control various plant diseases. Most of these drugs are absorbed by and translocated throughout the plant, providing systemic therapy. Streptomycin is used against a variety of bacterial pathogens, tetracycline is able to control the growth of certain mycoplasmas, and cycloheximides offer effective control for certain diseases caused by fungi.

## Host Resistance and Selection

Disease-resistant varieties of plants offer an effective, safe, and relatively inexpensive method of control for many crop diseases. Most available commercial varieties of crop plants bear resistance to at least one, and often several, pathogens. Resistant or immune varieties are critically important for low-value crops in which other controls are unavailable, or their expense makes them impractical. Much has been accomplished in developing disease-resistant varieties of field crops, vegetables, fruits, turf grasses, and ornamentals. Although great flexibility and potential for genetic change exist in most economically important plants, pathogens are also flexible. Sometimes, a new plant variety is developed that is highly susceptible to a previously unimportant pathogen.

## Variable Resistance

Resistance to disease varies among plants; it may be either total (a plant is immune to a specific pathogen) or partial (a plant is tolerant to a pathogen, suffering minimal injury). The two broad categories of resistance to plant diseases are vertical (specific) and horizontal (nonspecific). A plant variety that exhibits a high degree of resistance to a single race, or strain, of a pathogen is said to be vertically resistant; this ability usually is controlled by one or a few plant genes. Horizontal resistance, on the other hand, protects plant varieties against several strains of a pathogen, although the protection is not as complete. Horizontal resistance is more common and involves many genes.

## Obtaining Disease-resistant Plants

Several means of obtaining disease-resistant plants are commonly employed alone or in combination. These include introduction from an outside source, selection, and induced variation. All three may be used at different stages in a continuous process; for example, varieties free from injurious insects or plant diseases may be introduced for comparison with local varieties. The more promising lines or strains are then selected for further propagation, and they are further improved by promoting as much variation as possible through hybridization or special treatment. Finally, selection of the plants showing greatest promise takes place. Developing disease-resistant plants is a continuing process.

Special treatments for inducing gene changes include the application of mutation-inducing chemicals and irradiation with ultraviolet light and X-rays. These treatments commonly induce deleterious genetic changes, but, occasionally, beneficial ones also may occur.

Methods used in breeding plants for disease resistance are similar to those used in breeding for other characters except that two organisms are involved—the host plant and the pathogen. Thus, it is necessary to know as much as possible about the nature of inheritance of the resistant characters in the host plant and the existence of physiological races or strains of the pathogen.

## Use of Genetic Engineering in Developing Disease-resistant Plants

The techniques of genetic engineering can be used to manipulate the genetic material of a cell in order to produce a new characteristic in an organism. Genes from plants, microbes, and animals can be recombined (recombinant DNA) and introduced into the living cells of any of these organisms.

Genetically modified organisms that have had genes from other species inserted into their genome (the full complement of an organism's genes) are called transgenic. The production of pathogen-resistant transgenic plants has been achieved by this method; certain genes are inserted into the plant's genome that confer resistance to such pathogens as viruses, fungi, and insects. Transgenic plants that are tolerant to herbicides and that show improvements in other qualities also have been developed.

Apprehension about the release of transgenic plants into the environment exists, and measures to safeguard the application of this technology have been adopted. In the United States several federal agencies, such as the U.S. Department of Agriculture, the Food and Drug Administration, and the Environmental Protection Agency, regulate the use of genetically engineered organisms. As of 2016, more than 457 million acres (185 million hectares) worldwide were planted with genetically modified (GM) crops. Among the most successful GM crops are corn (maize), soybeans, and cotton, all of which have proved valuable to farmers with respect to producing increased yields and having economic advantages.

## Anthracnose

Anthracnose is a fungal disease that tends to attack plants in the spring when the weather is cool and wet, primarily on leaves and twigs. The fungi overwinter in dead twigs and fallen leaves. Cool, rainy weather creates perfect conditions for the spores to spread. Dry and hot weather stop the progression of the disease that may begin again once the weather conditions become optimal. The problem can be cyclic but is rarely fatal.

Anthracnose fungus infects many deciduous and evergreen trees and shrubs, as well as fruits, vegetables and grass. Anthracnose is noticeable along the leaves and the veins as small lesions. These dark, sunken lesions may also be found on stems, flowers and fruits.

In order to distinguish between anthracnose and other leaf spot diseases, you should carefully examine the undersides of leaves for a number of small tan to brown dots, about the size of a pin head.

A wide variety of plants can be affected by anthracnose fungus, including those grown outside of a greenhouse, such as woody ornamentals and tropical foliage plants.

Potted plants and greenhouse crops such as cyclamen, ficus, lupine, palms, succulents and yuccas are sometimes affected.

Trees and shrubs that are prone to anthracnose include maple, camellia, walnut, ash, azalea, oak and sycamore.

### Identification of Anthracnose

- On leaves, anthracnose generally appears first as small, irregular yellow or brown spots. These spots darken as they age and may also expand, covering the leaves.
- On vegetables, it can affect any part of the plant.

- On fruits, it produces small, dark, sunken spots, which may spread. In moist weather, pinkish spore masses form in the center of these spots. Eventually, the fruits will rot.
- On trees, it can kill the tips of young twigs. It also attacks the young leaves, which develop brown spots and patches. It can also cause defoliation of the tree.

Anthracnose can affect many plants with its brown spots, including this cucumber leaf.

### Control and Prevention

### Control Anthracnose

- Remove and destroy any infected plants in your garden. For trees, prune out the dead wood and destroy the infected leaves.
- You can try spraying your plants with a copper-based fungicide, though be careful because copper can build up to toxic levels in the soil for earthworms and microbes. For trees, try a dormant spray of bordeaux mix.

### Prevent Anthracnose

- Plant resistant plants, or buy healthy transplants.
- Plant your plants in well-drained soil. You can also enrich the soil with compost in order to help plants resist diseases.
- Water your plants with a drip sprinkler, as opposed to an overhead sprinkler. Don't touch the plants when they are wet.
- Keep ripening fruits from touching the soil.
- Remember to rotate your plants every 2 to 3 years.

## Apple Scab

Apple scab occurs everywhere in the world where apples are grown and results in more losses than any other apple disease. It is most serious in areas that have cool, wet spring weather and may not be economically important in warm and dry climates.

## Symptoms and Signs

Apple scab results in symptoms on most upper plant parts, most notably leaves and fruit. Petioles, flowers, sepals, pedicels, young shoots, and bud scales can also become infected.

## Apple Leaf and Fruit Symptoms

Apple scab infections are initiated in early spring on emerging and young leaves. Early lesions appear 10 days later as lighter green areas compared to the surrounding leaf tissue. Lesions increase in size and become olive-colored and velvety as a result of asexual spore production (conidia). Scab lesions that form on young leaves may expand to more than 1 cm in diameter. Ontogenetic resistance of older leaves, however, usually results in smaller lesions or no visible symptoms. Affected tissues eventually may become distorted and puckered, and leaf lesions become cracked and torn. Severely infected leaves drop from trees. Two to three consecutive defoliation events can weaken trees, resulting in a greater susceptibility to other stresses such as freeze damage, insect injury, and other diseases.

Fruit lesions are generally blistered and "scabby" in appearance, with a distinct margin. The earliest noticeable symptom on fruit is water-soaked areas that rapidly develop into velvety, green to olive-brown lesions. Infections of young fruit cause fruit distortion as healthy tissue continues to grow. Severely infected fruit often drop prematurely from trees.

Figure: Early lesions are olive-colored (A) and velvety with no distinct margins (B).

Figure: Fruit lesions appear blistered and scabby (A) with distinct margins (B).

Figure: Affected leaves become distorted and puckered.

Figure: Infections of young fruit cause distortion as fruit mature

## Apple Blossoms

Blossom infections usually develop into small, dark green lesions at the base of flowers, on sepals, and on stem pedicels before and during bloom. Developing fruit may drop from infected pedicels, resulting in lower fruit yield. Reduced and late-return bloom can result from infections that had occurred in preceding years.

## Pathogen Biology

Venturia inaequalis is an ascomycete fungus; it produces sexual spores (ascospores) in a sac-like structure called an ascus (plural asci). The mycelium of V. inaequalis is septate, and the nuclei are haploid.

## Sexual Reproduction

Venturia inaequalis overwinters in fallen leaf and fruit debris as pseudothecial initials. These structures emerge in early spring at approximately the same time as host plants break dormancy.

Mating between the two different mating types takes place in debris, and both mating types must be present in order for sexual reproduction to be initiated. Mating consists of the fusion of a male organ (antheridium) formed from a hyphal tip of one mate to a female receptive hypha (trichogyne) from the opposite mate. The trichogyne is attached to a coil of hyphae called the pseudothecial initial. During fertilization, nuclei pass from the antheridium through the trichogyne into a cell at the base of the pseudothecial initial.

After fertilization, the pseudothecial initial develops into a pseudothecium, a cavity located within a dense mat of fungal mycelia called a stroma. Inside this cavity, asci and ascospores are formed. The very brief diploid stage in the life cycle of V. inaequalis occurs within the pseudothecium in single hyphal cells (croziers), which give rise to the haploid ascospores following meiosis. Asci are elongated, sac-like structures, each of which contains eight ascospores in a linear arrangement. The ascospores are brown, two-celled spores, and have a characteristic "footprint" shape. The shape of ascospores inspired the Latin name for apple scab, "inaequalis", which refers to the unequal size of the cells. Ascospores measure between 5 and 7 μm wide and between 11 and 15 μm long.

During wet conditions, mature pseudothecia swell and protrude from the surface of fallen leaves. Ascospores are released and carried to blossoms and leaves by rain and wind. There is only one cycle of ascospore production and infection per season.

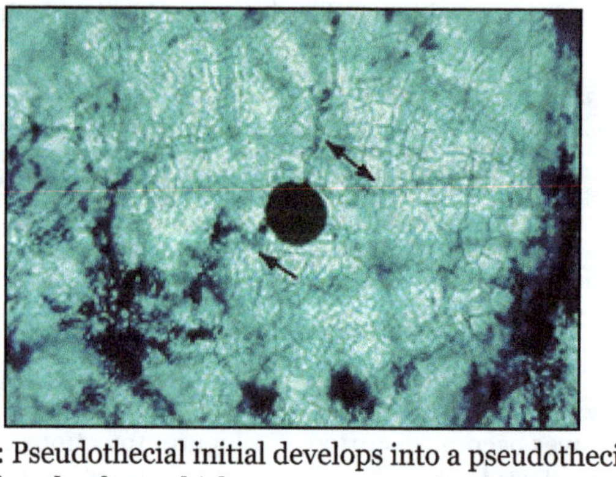

Figure: Pseudothecial initial develops into a pseudothecium.
Arrows show hyphae, which may represent the two mating types.

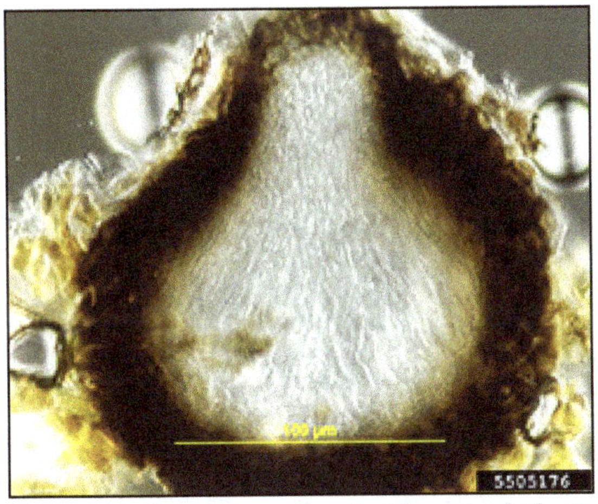

Figure: Asci and ascospores form inside a pseudothecium. Asci and ascospores may be stained (not shown) for enhanced viewing.

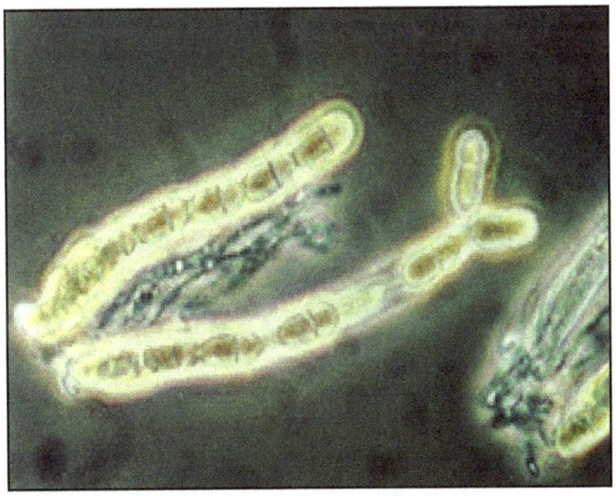

Figure: Asci of Venturia inaequalis containing eight ascospores. Arrows highlight individual ascospores.

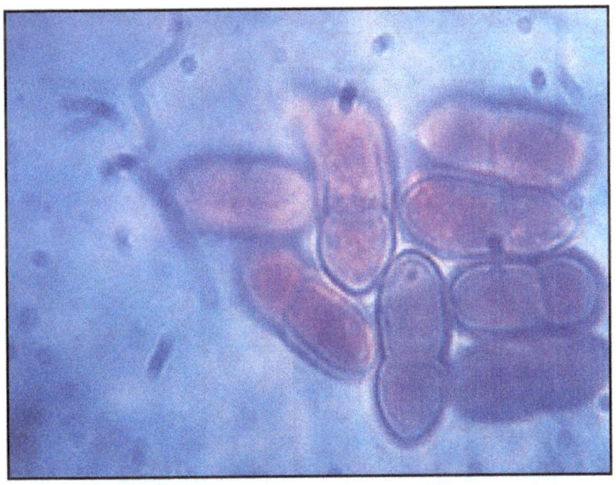

Figure: Two-celled ascospores of Venturia inaequalis are footprint-shaped.

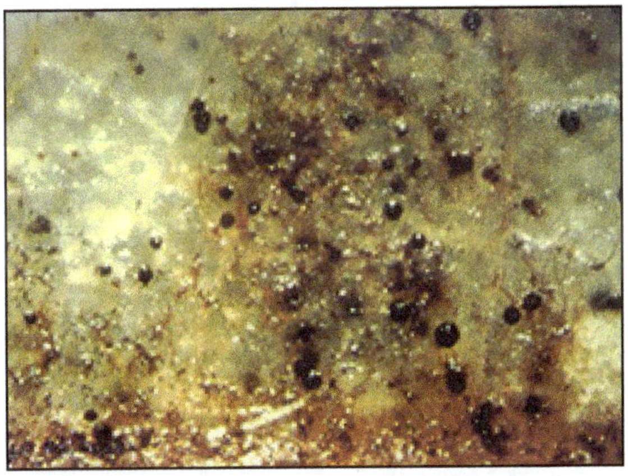

Figure: View of a fallen apple leaf, showing the mature pseudothecia of Venturia inaequalis protruding above the surface ready to release ascospores.

## Asexual Reproduction

V. inaequalis reproduces asexually by spores called conidia. Conidia are single-celled, uninucleate, and narrower at one end than the other. In mass, conidia appear brown or olive, but they are lighter when viewed individually under a microscope. Conidia measure between 6 and 12 µm wide and 12 and 22 µm long and are produced by specialized short hyphae called conidiophores. Conidiophores are formed on a dense mat of mycelia that pushes up through and ruptures the leaf cuticle. It is this mass of conidia and conidiophores that causes the velvety appearance of young scab lesions.

Conidia are produced nine to thirty days after initial leaf infection, depending upon temperature. They are disseminated by wind and by wind-driven rain. Both ascospores and conidia require a period of wetness in order to germinate. The germination hypha penetrates the cuticle and establishes a new infection. There can be many cycles of conidial production and infection within a single growing season.

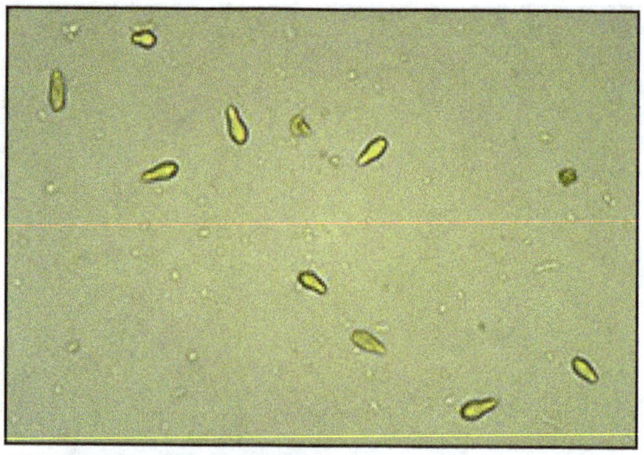

Figure: Conidia are asexual, single-celled spores.

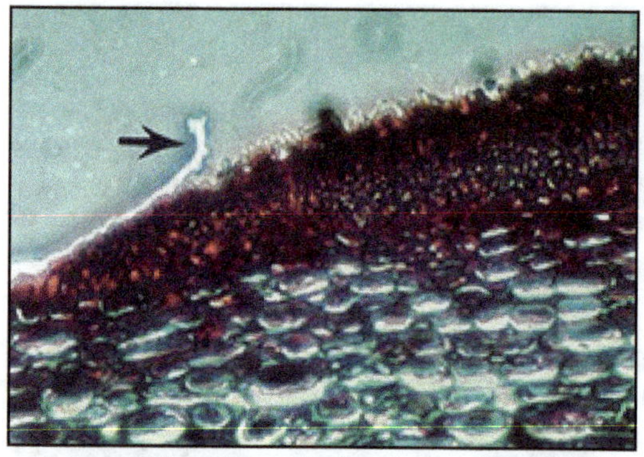

Figure: Cross section through a leaf infected with apple scab. Arrow indicates the leaf cuticle, which has been ruptured and pushed back by the mass of erupting conidia and conidiophores.

## Disease Cycle and Epidemiology

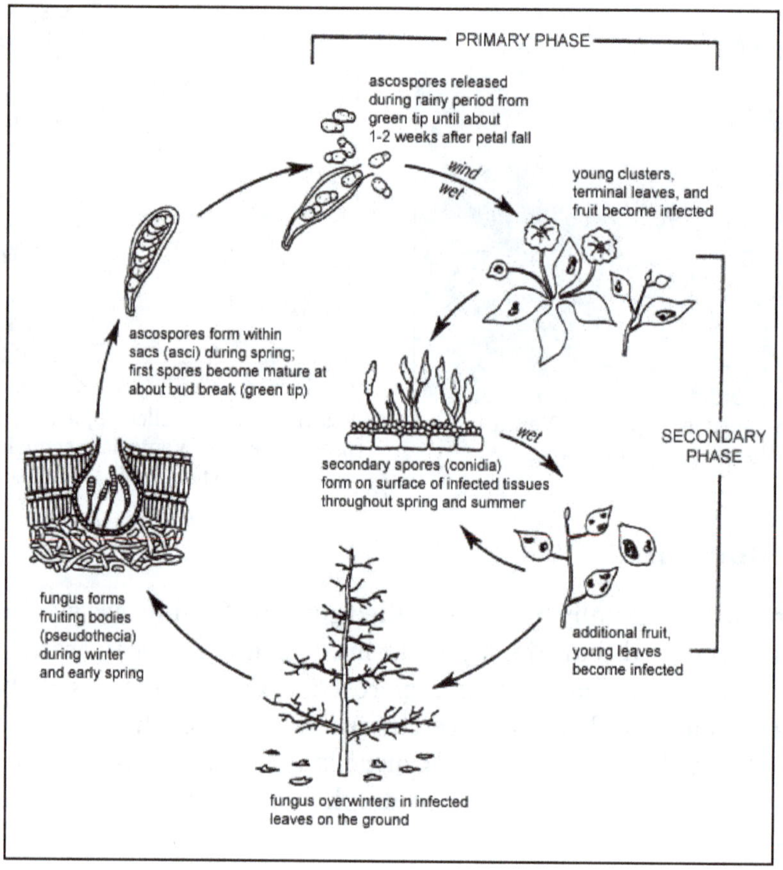


Figure: Disease cycle of apple scab.

## Disease Cycle

During the growing season, the apple scab fungus is found only in the area between the host cuticle and the epidermis. Once infected leaves fall from trees and die, the fungal mycelia colonize them completely. Venturia inaequalis usually overwinters in fallen leaves as mycelia and pseudothecial initials. In milder climates, the fungus may overwinter in infected buds and produce conidia that serve as primary inoculum. In leaf debris, most pseudothecial initials form within a month after leaves fall, and then enter a period of dormancy.

In early spring, mating (sexual reproduction) takes place in leaf debris between two mating types (as described above). In response to rain events, asci expand through the ostiole, a hole at the top of the pseudothecia. Ascospores are forcibly discharged and spread by wind and by wind-driven rain. Sufficient moisture must be present for ascospores to infect blossoms and young leaves. In most years, ascospore dissemination coincides with the several-week period between budbreak and the end of bloom.

Lesions caused by ascospores (primary infections) produce asexual conidia within 9 days and up to 30 days later. Conidia are dispersed to healthy leaves and developing fruit, where they establish secondary infections. Up to 100,000 conidia can be produced by a single lesion. Lesions from primary or secondary infections expand at a rate that is determined partly by temperature and partly by characteristics of the host tissue, including genotype and age. Lesion expansion, in turn, affects the rate at which new spores are produced. During cooler conditions, or on more resistant cultivars, lesions expand more slowly and may be smaller in size. As a result, the number of secondary cycles will be fewer.

Infection and spore production are dependent upon available moisture. Infection by ascospores and conidia is highly dependent upon how long the leaves or fruit stay wet, as well as on the average temperature. The Mills table relates leaf wetness duration and temperature to determine the likelihood that conidial infection will occur. For example, at an average temperature of 18° C (65° F), light infection will result if leaves remain wet for 9 h. Lesions can produce conidia after 9 days if the temperature averages 18° C (65° F), but not until 17 days if the temperatures are lower, averaging only 8° C (49° F). The Mills table continues to be revised as more data are gathered from different regions, and modifications by A.L. Jones comprise the most commonly accepted version.

Table: Wetting period (in hours) required for apple scab infection at different air temperatures, and time required for development of conidia by lesions at different air temperatures.

| Wetting period (HOURS) | | | | | |
|---|---|---|---|---|---|
| Average Temperature (F) | Average Temperature (C) | Light Infection | Moderate Infection | Heavy Infection | Incubation Period (days) |
| 78 | 25.6 | 13 | 17 | 26 | ... |
| 77 | 25.0 | 11 | 14 | 21 | ... |
| 76 | 24.4 | 9.5 | 12 | 19 | ... |
| 63-75 | 17.2-23.9 | 9 | 12 | 18 | 9 |
| 62 | 16.7 | 9 | 12 | 19 | 10 |
| 61 | 16.1 | 9 | 13 | 20 | 10 |
| 60 | 15.6 | 9.5 | 13 | 20 | 11 |
| 59 | 15.0 | 10 | 13 | 21 | 12 |
| 58 | 14.4 | 10 | 14 | 21 | 12 |
| 57 | 13.9 | 10 | 14 | 22 | 13 |

| | | | | | |
|---|---|---|---|---|---|
| 56 | 13.3 | 11 | 15 | 22 | 13 |
| 55 | 12.8 | 11 | 16 | 24 | 14 |
| 54 | 12.2 | 11.5 | 16 | 24 | 14 |
| 53 | 11.7 | 12 | 17 | 25 | 15 |
| 52 | 11.1 | 12 | 18 | 26 | 15 |
| 51 | 10.6 | 13 | 18 | 27 | 16 |
| 50 | 10.0 | 14 | 19 | 29 | 16 |
| 49 | 9.4 | 14.5 | 20 | 30 | 17 |
| 48 | 8.9 | 15 | 20 | 30 | 17 |
| 47 | 8.3 | 17 | 23 | 35 | 17 |
| 46 | 7.8 | 19 | 25 | 38 | 17 |
| 45 | 7.2 | 20 | 27 | 41 | 17 |
| 44 | 6.6 | 22 | 30 | 45 | 17 |
| 43 | 6.1 | 25 | 34 | 51 | 17 |
| 42 | 5.5 | 30 | 40 | 60 | 17 |

## Disease Management

### Chemical Management

Management of apple scab is focused on the prevention of primary infection by ascospores. This is because early infection by ascospores may result in poor fruit set and will result in more secondary inoculum throughout the season. Fungicide applications are therefore timed to coincide with the spring release of ascospores (between bud break and petal fall). Maturing plant tissue is somewhat resistant to infections, and scab-targeted fungicides are reduced or sometimes eliminated from spray schedules unless conditions are conducive for disease (wet weather, susceptible cultivars, and high disease pressure). Later-season fungicide sprays are often targeted toward other fungal diseases, but also may be effective against secondary inoculum. Disease severity and distribution varies, so growers should utilize local university sources when developing spray schedules.

Figure: Contact fungicides require complete, uniform coverage. Leaf appears to be completely covered with residue.

Figure: An airblast sprayer being used to apply fungicides in a commercial orchard.

Various fungicide chemistries are available for management of apple scab. Most products can be classified as contact or systemic. Contact fungicides form an external barrier on plant surfaces. They are typically not rain fast (they wash off readily with rain) and do not translocate through plant tissue. Systemic fungicides, in contrast, penetrate epidermal tissue and can translocate within plant tissue. They are more rain fast than contact fungicides, and they usually have longer reapplication intervals.

Both contact and systemic fungicides are applied before infection to prevent fungal spores from germinating or penetrating host tissue. Fungicides do not cure disease, even though some older recommendations may refer to systemic fungicides as curative products. Thorough spray coverage and uniform deposition are essential, especially for contact fungicides. To ensure coverage of newly emerging tissues and due to breakdown of active ingredients, fungicides must be re-applied on a regular schedule. This application schedule is dependent upon product formulation and systemic properties; consult product label for individual mrecommendations. In commercial orchards, air blast sprayers are typically used to apply fungicides to ensure sufficient coverage.

Fungicides are classified according to their mode of action. The Fungicide Resistance Action Committee (FRAC) has categorized fungicides according to their mode of action and, accordingly, their risk for resistance development. Many newer fungicides have a high risk for resistance, so it is important to rotate between FRAC groups to prevent development of resistant populations.

Timing of fungicide applications is critical for effective scab management. When fungicides are combined with cultural practices and are based on environmental conditions (e.g. prediction models such as a Mills table-based weather monitoring system) disease is more effectively managed. Fungicide applications based on weather monitoring systems can reduce numbers of sprays and/or more accurately time applications, thereby reducing inputs and increasing grower profits. Several university programs and private industries have developed regionally-specific prediction models that utilize weather forecasts to estimate leaf wetness and temperature conditions to evaluate risk for spore release and infection.

## Biological Control

There are a limited number of biological control products available for management of apple scab. To date, many of these products have below-average efficacy ratings when compared to traditional fungicides; thus, use of cultural controls and resistant cultivars are vital components of a biological control disease management program.

## Genetic Resistance

Scab-resistant (or more appropriately scab-tolerant) apples and crabapples are widely available. Levels of resistance/tolerance vary with cultivar, although most scab-resistant apple cultivars carry the Vf resistance gene from M. floribunda 821. For many years, resistance resulted in reduction and elimination of fungicide use for scab management.

In many areas, resistance remains intact. Risk for loss of host resistance (due to adaptability of the pathogen) is high when growers rely on resistance alone.

In all cases, an integrated management (IPM) program helps prevent resistance breakdown, as well as manage disease in orchards where resistance has been compromised. Cultural practices (improved air circulation and sanitation) are critical for reduction of inoculum. Additionally, use of fungicides (synthetic, organic, and biological) during primary infection periods also reduce inoculum and help break the disease cycle. Resistant and susceptible cultivars should not be planted in close proximity. Integrated use of these and other management practices help protect resistance and reduce disease pressure. Resistant or tolerant apple cultivars are gaining acceptance by both growers and consumers. Resistant cultivars are listed in table below.

Table: List of popular scab-resistant apple cultivars.

**Resistant Apple Cultivars**

Include but are not limited to:

- CrimsonCrisp
- Dayton
- Enterprise
- Freedom
- Galarina
- Gold Rush
- Jonafree
- Liberty
- Mac-free
- Novamac
- Prima
- Pricilla
- Pristine
- Redfree
- Sir Prize
- William's Pride

Susceptible cultivars are still widely grown due to popularity and consumer demand. Additionally, these cultivars often can be grown easily in drier regions where apple scab is not a problem. Cultivars that are highly susceptible are listed in table below.

Table: List of popular apple cultivars that are susceptible to scab.

**Susceptible Apple Cultivars**

Include but are not limited to:

- Blushing Gold
- Cortland
- Empire
- Fuji
- Gala
- Ginger Gold
- Golden Delicious
- Granny Smith
- Idared
- Jerseymac
- Jonathan
- Lodi
- McIntosh
- Mutsu
- Pink Lady
- Red Delicious
- Rome Beauty
- Stayman Winesap
- Winesap
- York Imperial

Scab-resistant flowering crabapples provide greater bloom potential because there is no risk for defoliation. Many scab-resistant cultivars are available, and many new ones are being introduced. Table below includes examples of crabapples from the Morton Arboretum that exhibit excellent resistance.

Table: List of popular scab-resistant crabapple on display at the Morton Arboretum near Chicago, IL.

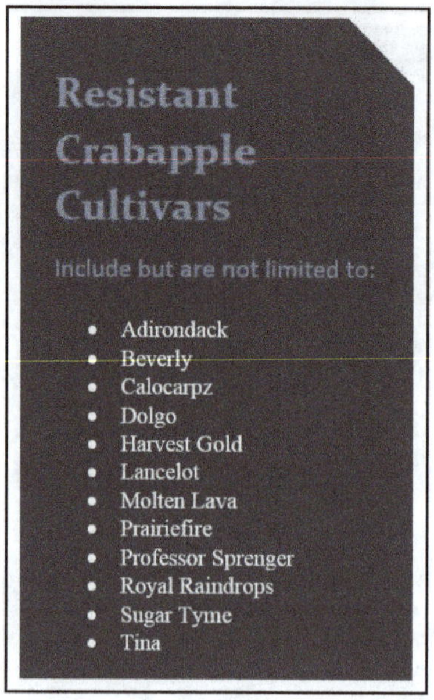

Resistant Crabapple Cultivars

Include but are not limited to:

- Adirondack
- Beverly
- Calocarpz
- Dolgo
- Harvest Gold
- Lancelot
- Molten Lava
- Prairiefire
- Professor Sprenger
- Royal Raindrops
- Sugar Tyme
- Tina

## Cultural Practices

Cultural practices should be combined with fungicides (chemical and biological) for an integrated pest management system. Two common cultural practices used to reduce apple scab are sanitation and increased air circulation. Sanitation eliminates inoculum during the growing season and during the overwintering season. Commercial growers use various methods to speed up leaf decomposition, thereby decreasing the amount of primary inoculum the following season. These include: applying urea to trees, just before leaf drop, applying urea to fallen leaves, and tilling fallen leaves into the soil or chopping them into small pieces. Another effective cultural practice is to lower wetness and humidity in tree canopies so that the environment becomes unfavorable for disease. Pruning and wide tree spacing can enhance air movement and allow sunlight to penetrate, which speeds up drying of leaves and fruit.

## Significance

Apple scab has long been a problem on apples; symptoms of the disease can be recognized on fruit in paintings from the fifteenth and sixteenth centuries, as well as the 1824 painting by American James Peale, shown in figure below.

Figure: Fruit Still Life with Chinese Export Basket.

The frequent historical depictions of scab-infected apples suggest that the disease was common and that the affected fruit was acceptable in earlier times. All the commonly grown apple cultivars were susceptible to the disease, and there were no chemical treatments until the late 1800s. At that time, copper- and sulfur-based fungicides served as protectants, but the treatments often caused substantial damage to the apple foliage.

## Bacterial Canker

Bacterial canker on trees is caused by the bacteria Pseudomonas syringae, a relatively weak pathogen that regularly coexists with host trees without anyone noticing. When this infection does become symptomatic, usually on stone fruit trees, the bark on affected limbs turns brown and those limbs or trees may refuse to bloom or leaf out in the spring. Sometimes, trees will leaf out despite a raging case of bacterial canker, but these new leaves quickly wilt and die.

### Symptoms

You may see the following symptoms:

- On stems and spurs: Sunken, dead areas of bark develop in spring and early summer, often accompanied by a gummy ooze. If the infection spreads all-round the branch it will die rapidly. However, it should be noted that gum production (gummosis) from the bark of Prunus species is actually quite common, and in the absence of dead, sunken bark is likely to have resulted from causes other than bacterial canker, for example physical damage or environmental stresses.
- On leaves: Small brown spots appear which are often round and fall out later to leave holes – as if the leaf had been hit by shotgun pellets, leading to the popular name of 'shothole'.

### Conditions for Disease Development

The bacterium enters the plant via wounds and stomata. Clavibacter may be seed-borne and may infest the seed externally or under the seed coat. High relative humidity and daytime temperatures between 25° and 30° C (77° and 86° F) generally favor the disease. Dense plant populations and overhead irrigation also provide an ideal environment to spread the bacterium. Insects, tools and human contact may also aid the spread.

### Bacterial Canker Control

Treatment of bacterial canker is generally mechanical, with the infected branches being removed using sterile pruning tools. Wait until late winter, if at all possible, and cauterize the wound with a hand-held propane torch to prevent reinfection by bacterial canker. It may be easier to see where the cankers end if you peel the bark back on infected branches until you find clean, white flesh. Cut at least 6 inches beyond the wound nearest the tree, in case the infection is spreading inward. Disposing of infected tissues immediately can also help stop the spread of this disease.

Once bacterial canker is removed from your trees, prevention is paramount. Often, trees become seriously infected when they're under stress or lack the proper nutrients. Trees planted in hard soils or where they cannot root deeply are most at risk. Check the pH of the soil around your tree and gently amend the top 16 inches with lime if needed. Foliar sprays of micro-nutrients, including zinc and boron seem to be protective, especially if applied in the fall or spring.

Nematodes can increase tree stress levels – if you have to replant, carefully fumigate your chosen site and select species grafted onto the most nematode resistant rootstocks you can find — huardian rootstocks are known to be protective against these tiny roundworms.

## Black Knot

Black knot is a widespread fungal disease that attacks plum and cherry trees, both fruiting and ornamental. The fungus, Apiosporina morbosa, (also identified as Dibotryon morbosum and Plowrightia morbsum), singles out trees of the genus prunus, which includes peach, apricot, and chokecherry. Once established, black knot is easily identified with its hard, uneven, black galls that seem to enwrap twigs and branches. Black knot is a slow developer, taking a season before its visually apparent and producing spores. The trick to controlling the fungus is identifying the infection well-before the disease becomes firmly established. If left to grow, it effectively strangles new growth, girdling branches and dooming the tree to deterioration and poor fruit production. Insects and plant diseases use the galls as an entry to the tree.

The disease cycle starts when spores are released from established knots where the fungus overwinters. This occurs during damp spring conditions when temperatures reach 60 degrees or higher. The spores travel to other parts of the tree and, depending on the breezes, to nearby host trees. The spores germinate on stems beneath a thin film of moisture, often at the juncture of a new leaf start. They form small, olive-colored swellings over the first season, darkening in color as the season progresses, hidden by the leaves they'll eventually kill,. By the second year, the galls are expanding quickly, especially where the weather remains humid. The growing infection begins releasing its own spores as it swells into the dark, easy-to-spot (especially after leaves have fallen) warty black fungus that coils along stems and branches. At this point, astute pruning and chemical treatments may not be enough to save the tree, no matter how careful the pruner is not to spread spores or leave them behind when removing the galls.

Paying close attention to your fruit trees and catching the infections as soon as they're apparent, followed by quick pruning and careful disposal of the gall-infected branches, can save trees. Organic treatments can also help protect trees while keeping harmful chemicals off your fruit.

When choosing new plantings, consider that some varieties of tree and shrubs are more susceptible to the disease than others. They should be avoided in areas where the fungus is prolific. Tart cherry varieties are said to be less susceptible to the disease than sweet. Japanese plums are said to be less susceptible than American varieties. A number of plums, including President, Early Italian, Santa Rosa and Shiro carry varying degrees of resistance to the fungus. Susceptibility varies depending on the climate zones. Varieties that are susceptible in humid southern climates may be less so in dryer or cooler ones. Talk to your local nursery staff to see which varieties of plums, cherries, and ornamentals do best in your area.

## How to Control

- Inspect your trees carefully for first signs of the disease. This is best done in winter, when leaves are absent, but should be continued as well throughout the growing season. Look for cracks, discoloration, swelling, or other first signs of infection. Check carefully around twig and leaf axils.
- Remove any knots that are found. This is best done during winter when spore production is down. Cut well-past the galls, four to eight inches, to ensure all the infection and its spores are removed. Larger branches with established knots should be removed entirely. Use a pruning knife or chisel to remove galls on trunks and large branches, cutting down to the wood and out to at least an inch beyond the infection.
- Continue to inspect for and remove galls as the season progresses.
- Take care not to spread spores when pruning trees with black knot. Don't allow twigs or other cuttings to fall to the ground where the spores could survive.
- Dispose of infected stems and branches by burying or, where allowed, burning. Small cuttings can be stuffed in trash bags and hauled away. Do not compost any infected cuttings unless your heap has an internal temperature of 160 degrees (not many do).
- Clean pruning tools as you use them with a solution of 1/2 cup bleach to a gallon of water. Wipe tools between cuts and leave your pruning blades in the solution for three to six minutes when finished. Or use a safe, commercial fungicide cleaner such as Physan 20.
- The Alberta Ministry of Agriculture and Forestry has a detailed guide to pruning and disposing of infected trimmings at their website.
- Fungicides can offer significant protection against black knot, but are unlikely to be effective if pruning and sanitation are ignored. Organic gardeners will want to avoid all but OMRI listed fungicides.
- Spraying trees with NEEM oil, a natural fungicide that controls leaf spot, rust, scab, and other tree fungus, can help inhibit the spread of black knot (it will not kill fungus that is already present). Spray trees per instructions just ahead of leaf and blossom emergence and, if possible, ahead of rain. Continue on a 7-10-day cycle until weather dries. Use of other fungicides can also discourage spores from germination. But few are specifically indicated for use on already infected trees.
- Spraying lime sulfur on trees during the dormant period is said to prevent the production of spores. Copper sprays applied during dormancy may also inhibit spore production.
- Take out wild cherry and plum trees around your property. They harbor the disease and release spores that are easily carried to your susceptible nursery trees.
- When planting new trees, place them away and upwind from established or wild prune and cherry trees.

## Blossom-end Rot

Blossom end rot is a disorder related to plant nutrition that creates a dark, sunken area on the blossom end (bottom) of a fruit. It is a common issue with many fruiting plants, including tomatoes, eggplant, peppers and more. Blossom end rot is often related to a crop's lack of calcium and can be prevented, but not corrected.

### How to Stop Blossom End Rot

Blossom-end rot is a disorder caused by calcium deficiency in the plant.

While this may be a result of low calcium levels in the soil, more often than not, it is the result of erratic watering.

When the plant is allowed to get too dry, or is given too much water over a period of time, its ability to absorb calcium from the soil is greatly diminished. In other words, your soil might have plenty of calcium, but it just isn't bioavailable and being taken up and utilized by the plant.

If your soil is indeed low in calcium (determined by a soil test) the easiest solution is to add garden lime several times per year, according to the directions on your soil test results. (Don't just add lime without testing your soil first, as you might disrupt the optimal pH for growing your crops).

Over fertilization, especially with high nitrogen fertilizer, can also cause blossom-end rot. Over fertilization can cause such rapid growth that nutrients such as calcium won't be able to keep up with the growth. Always soil test before fertilization and fertilize according to the results.

You can also choose varieties of tomato that are resistant to blossom-end rot. Varieties of tomatoes that produce large fruit tend to be more susceptible to blossom-end rot, while varieties with smaller fruit are less likely to suffer it.

Blossom-end rot is much easier to prevent than it is to cure. Once it has set in, it can be really hard to reverse, but there are a few things you can do that have a good chance of turning things around.

### Ensure Consistent Water Supply

If the issue is erratic moisture, here are some tips:

1. Make sure your soil isn't allowed to dry out. The best defense against blossom end rot is a nice, consistent soil moisture level.
2. As the summer rolls on, it is easy to forget to water the garden regularly. If it is hard for you to be consistent, or if you plan to take a vacation, consider installing a drip irrigation system and a timer. Drip irrigation is a major time-saver, affordable, water-wise and an outstanding disease prevention method because it both keeps soil moisture regular and keeps water (and therefore fungus spores) from getting onto the leaves of your crops.
3. Mulch, by adding a three-inch layer of organic mulch, you can help maintain adequate soil moisture levels, even during dry spells. It is best to add the mulch after your soil has warmed in the spring.

4. Plant susceptible crops (such as tomatoes, melons, squash, peppers, and eggplants) in well-drained soil that has been amended with compost or well-rotted manure. Soil amended with plenty of organic matter will retain moisture better and supply plenty of nutrition (including calcium) to your plants.

### Fortify your Crops at Planting

In addition to making sure you have consistent moisture levels in your soil, you can fortify your plants when you put them in the ground to make sure they get plenty of calcium throughout the season.

Many people use garden lime to adjust their garden pH and add calcium at the time of planting. This will treat the entire garden soil. (If your soil pH doesn't need adjusting, use gypsum instead of lime.) You can also add 2-3 Tums tablets or other calcium carbonate antacid to each planting hole to add extra calcium.

If you already have signs of blossom-end rot, you can make a solution from 2-3 calcium carbonate antacid tablets, 8 ounces of milk and a quart of pure water, and irrigate your plants with it daily to help keep blossom-end rot from destroying more of your crops than it has to. This might help turn things around, but isn't foolproof.

Don't bother with the calcium sprays at the garden store that promise to stop blossom end rot. While they can help with other issues related to nutrient deficiency, to stop blossom end rot, the calcium has to come up from the soil through the roots, not through the leaves.

## Brown Rot

Brown rot is a fungal disease that commonly affects stone-fruit trees like peaches and cherries. Brown rot affects the fruit tree's flowers and fruit crop. Healthy trees can survive it if they end up with brown rot, but it's best to control this disease as soon as possible. Fortunately, brown rot is easy to spot, prevent, and treat once you know what to look for and how to keep it from thriving in your fruit garden.

What does brown rot look like what are the visible signs

- Brown, wilted blossoms (common site of infection).
- Dark, sunken spots on new shoots and brown, hanging leaves on infected limbs.
- Affected fruit develops small spots of rot that enlarge quickly.
- Rotted fruit develops fuzzy tan/grey spores that cover the fruit surface.
- If left on the tree, fruit shrivels, darkens, and hardens into "mummies".
- If mummies are left on the tree or on the ground around the tree, fungal cycle is able to continue and recontamination occurs.

## Which Fruit Trees are Affected by Brown Rot

### Stone Fruit Trees

- Apricots (and Apriums)
- Cherries (both Sweet and Sour types)
- Nectarines
- Peaches
- Plums (and pluots/plumcots)

Occasionally, brown rot is also thought to affect Apples and Pears, but this is not common.

## What can be done to Prevent Brown Rot

Plant resistant varieties:

- Hardired Nectarine
- Harglow Apricot
- Venture Peach

Keep a clean, maintained growing site:

- Plant in a well-drained location with full sun (6-8 hours during the growing season).
- Prune regularly to keep trees open to light and air circulation (fights fungi).
- Remove any pruning or landscaping debris to avoid sites for fungi to thrive.
- Remove any damage or diseased fruit and limbs. These are more susceptible to infection.
- Thin fruit. Fruits that touch may pose disease risk.

## What can be done to Treat Brown Rot in Fruit Trees

### Proper Sanitation

- Prune out all signs of disease in limbs as soon as they appear. Clean pruners between cuts.
- Dispose of pruning and other debris to avoid recontamination – do not try to compost.
- Remove and dispose of all infected fruit. Do not leave them on the ground or try to compost.

### Fungicide Spray

- A wettable powder fungicide, a liquid concentrate fungicide, or natural copper-based fungicide spray/dust.
- Spray preventatively if brown rot is problematic in your areas, even before symptoms appear.
- Be sure the fungicide spray is recommended for use on the trees being sprayed (check label).
- Most spray applications start at pink-bud stage in spring and are repeated as needed.
- Follow instructions on the product label for specific application timing and frequency.

# Cedar-apple Rust

Cedar-apple rust is a common disease of apple and crabapple. The fungus that causes the disease, Gymnosporangium juniperi-virginianae, is unusual because it must spend a phase of its life cycle as aparasite on Juniperus species, such as red cedar or ornamental junipers.

Cedar-apple rust can be severe on apple. Infections of apple fruit result in lower fruit quality and early fruit drop. Leaf spots may cause early defoliation, especially during dry summers. If trees are defoliated several years in a row, they become weakened and unthrifty. Fruit bud formation may be reduced after one year. The disease is not as harmful to juniper, causing galls but not severely affecting plant vigor.

## Disease Cycle

The rust organism spends one full year of its life cycle on junipers. During the second spring, usually around the time crabapples are in bloom, the galls become rain soaked and swell, producing jelly-like tendrils (spore horns) that project out of the galls. As the spore horns begin to dry, the spores are released and carried by the wind to young, newly developing leaves of hawthorns and other susceptible plants. Dispersal of spores can range up to 5 miles from a juniper but most infections develop within several hundred feet. About a month after crabapples have bloomed, the spores are exhausted and most leaves are no longer susceptible. Ten-to-14-days from initial infection, small yellow spots can be seen on upper surfaces of infected leaves. Several weeks later, the fungus appears as orange or brown spots with hairlike appendages on the underside of the leaf. In late summer, the rust spots release the spores and are carried to nearby junipers.

Cedar-apple rust is the most common of the three fungal rust diseases and attacks susceptible cultivars of apples and crabapples. It infects the leaves, fruit, and, occasionally, young twigs. The alternate host plant, Eastern red cedar (Juniperus virginiana), is necessary for the survival of the fungus. Spores produced on rose family plants only infect juniper plants, and those originating on the evergreen host only infect rose family plants. Repeated infections of cedar-apple rust can be unsightly and seriously weaken and destroy the ornamental value and health of susceptible plants.

## Symptoms on Apple and Crabapple

On leaves: Bright yellow/orange spots develop on the upper surface of the leaves in late spring. These spots gradually enlarge, becoming evident on the undersurface of the leaves as small bulges. In midsummer, these rust lesions develop hair like, cylindrical tubes (hyphae), which release spores into the air that are blown to the juniper host. Infected leaves of apples and crabapples may drop, with defoliation more severe in dry summers. Galls that form on the juniper host do not become evident until July the next year, requiring two years for the fungus to complete its life cycle.

On twigs: The rust appears as a swollen corky gall on the current year's growth, usually no more than 1 inch in length. The swelling eventually develops the characteristic cylindrical fruiting bodies. Seriously affected twigs are stunted and may die.

On fruit: The rust causes yellow to orange spots similar to those found on the leaves, but the spots are usually much larger. Fruit infection causes an inferior fruit quality or premature fruit drop.

## Symptoms on Juniper

In mid-spring, swellings or galls develop on juniper needles that were infected with spores during the previous year. These galls are brown to dull red in color, globular in shape, and may vary from pea-sized to an inch or more in diameter. As they mature, circular pits or depressions are found over the surface of the galls. After spring rains and damp weather, yellow gelatinous tendrils or spore horns form in these pitted areas. The tendrils elongate rapidly and release spores during dry, windy weather that follows the spring rains. Spores produced on the juniper host are then blown to the apple, crabapple, and hawthorn hosts as their new growth emerges.

Eventually the galls dry out but remain attached to the tree for several years, resulting in some small twig and tip dieback.

## Control

Because apples are an economically important crop, control is usually focused there. Interruption of the disease cycle is the only effective method for control of the cedar apple rust. The recommended method of control is to "remove cedars located within a 1 mile (1.6 km) radius" of the apples to interrupt the disease cycle, though this method is seldom practical. For those doing bonsai, it is common to have the trees within feet of each other and on the central eastern seaboard of the United States, Eastern Red Cedar is a common first-growth conifer along roadsides.

There are differences in the susceptibility of various apple varieties. 'Jonathan', 'Rome Beauty', 'Wealthy' and 'York Imperial' are susceptible. 'Grimes Golden', 'Narragansett', 'Red Delicious', 'Winesap', 'Staymans', 'Redfree', 'Jonafree' and 'Priscilla' are resistant. Crabapples are generally more susceptible than apples. Resistant crabapples include 'Adams', 'Beverly',

'Candied Apple', 'Dolgo', 'Donald Wyman', 'Eleyi', 'Inglis', 'Indian Summer', 'Liset', 'Mt. Arbor', *M. persicifolia*, 'Red Jewel', 'Robinson', 'Robusta', 'Royalty', *M. sargentii*, 'Tina', 'Snowdrift', and 'Special Radiant'. Resistant *Crataegus* (Hawthorn) include *C. crus-galli*, series *Intricatae*, *C. laevigata*, 'Autumn Glory', *C. phaenopyrum*, *C. pruinosa*, *C. viridis*, and 'Winter King'. The resistant varieties are less susceptible to attack, but that does not mean that they are free from an aggressive attack.

Fungicide sprays applied in a timely manner are highly effective against the rust diseases during the apple cycle. Most sprays are applied four times at 7- to 10-day intervals, starting with pink bud on crabapples. These applications are to protect the apples from spores being released from the cedar host in mid-spring. If cedar apple rust disease is diagnosed on apple fruits and leaves it is far too late to spray. Systemic fungicides are available.

Fungicides listed for use on apples can be used in July and August on the cedars to reduce infection. Application of fungicides to the junipers before and while they are in the infectious orange gelatinous state seems to reduce the severity of the outbreak.

# Club Root

Club root is an infection of the roots of brassicas and related plants by Plasmodiophora brassicae, a soil-dwelling micro-organism related to the slime moulds, leading to massive swelling, distortion and severely retarded growth.

It affects Brussels sprouts, cabbages, cauliflowers, turnips, swedes and radishes, and ornamental relatives such as Cheiranthus (wallflowers), Matthiola (stocks), Aubrieta (aubretia) and cabbage-family weeds such as Capsella bursa-pastoris (shepherd's purse).

Club root can infect whenever the soil is moist and warm, so most new infections occur from mid-summer until late autumn.

## Symptoms

You may see the following symptoms:

- Above ground: Stunted growth, purplish foliage and wilting in hot weather, which may recover under wetter conditions.
- Below ground: The root system becomes massively swollen and distorted, with a loss of the finer roots.
- Growth and yield are severely reduced and very badly affected plants may die.

## Control

## Non-chemical Control

- If you buy brassica plants, take great care that they come from a guaranteed club root-free source. Be particularly careful in accepting plants from gardening friends, who with the best of intentions may be an unwitting source of infection.

- If the disease is known to be present, try to give plants a head start by growing them on in healthy soil to a larger than normal size before planting out, or as transplants in pots of at least 9cm (3½ inch) diameter, so that they begin growth in the affected soil with a larger than usual healthy root system.
- Beware of spreading contaminated soil on tools, wheelbarrows or footwear.
- Club root is reduced (but not eliminated) by raising the soil pH by liming. On acid soils, lime at the rate of 500g per sq m (15oz per sq yd), with lighter dressings of 270g per sq m (8oz per sq yd) in future years.
- Along with the liming regime, take care to improve drainage, by making raised beds for example.
- Keep down susceptible weeds like shepherd's purse, charlock, wild radish.
- Some cultivars show some levels of resistance: Brussels sprouts 'Cryptus F1' and 'Crispus F1', calabrese 'Monclano F1', cauliflower 'Clapton F1' and 'Clarify F1', cabbages 'Kilaxy F1', 'Kilazol F1' and 'Kilaton F1', kale 'Tall Green Curled' and swede 'Marian' also show resistance.

### Biology

Plasmodiophora brassicae is loosely described as a fungus but is in fact more closely related to the slime moulds. It produces resting spores that can contaminate soil for up to 20 years. In the presence of susceptible plant roots, these resting spores germinate and infect the root hairs, causing the distortion. The fungus produces more resting spores in the affected tissue, which eventually rots and releases them back to the soil, ready for the cycle to start over again.

## Corn Smut

Corn smut is caused by a fungus called Ustilago zeae, which is commonly blown on the wind from an infected stand to an uninfected stand of corn. The spores can live up to three years, making it very hard to completely destroy. The fungus is generally considered an opportunistic fungus, only able to move into the tissues of your corn plants through damaged or torn tissues, but if they get a chance to infect, they waste no time.

Once the Ustilago zeae spores find an opening in your corn, it takes about 10 days for the galls to appear. These unsightly growths vary in size, but can reach up to five inches across, with the smaller galls appearing on leaf and silk tissues and the larger erupting from maturing ears.

Although this fungus wasn't what you planted or even hoped for when you were thinking about growing corn, it's considered a delicacy in and of itself, as long as you harvest the corn smut galls while they're still young. In Mexico, they call it cuitlacoche and it's used in cooking in a similar way as a white mushroom.

## Treating Corn Smut Disease

Corn smut control can be difficult, if not impossible, to eliminate, but you can at least minimize the exposure your corn gets to the fungus year after year. Always make sure to clean up all the corn debris in your patch as it falls, since it can harbor more corn smut spores. If you remove the galls while they're still young, that will also help to reduce the spore exposure level.

If you've had corn smut problems in the past, trying a more resistant variety of sweet corn may help, too. Look for the white corn varieties before your next corn planting. These include:

- Argent
- Brilliant
- Fantasia
- Pristine
- Seneca Sensation
- Seneca Snow
- Prince Seneca
- Sugar Prince
- Silver King
- Silver Prince
- Summer Flavor 72W

## Biology

## Life Cycle

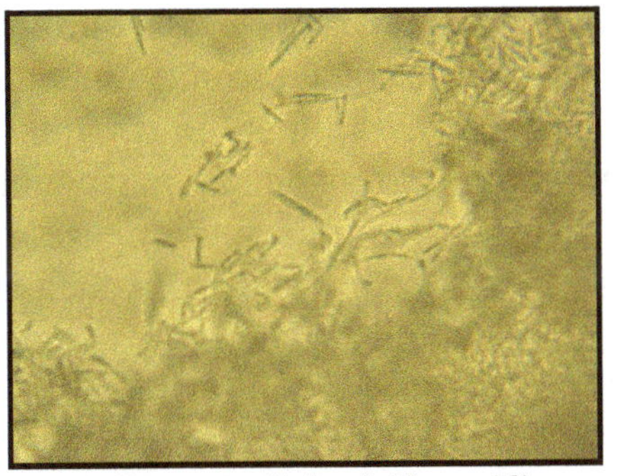

Ustilago maydis haploid sporidia.

When grown in the lab on very simple media, it behaves like baker's yeast, forming single cells called sporidia. These cells multiply by budding off daughter cells. When two compatible sporidia meet on the surface of the plant, they switch to a different mode of growth. First, they send out conjugation tubes to find each other, after which they fuse and make a hypha to enter the maize plant. Hyphae growing in the plant are dikaryotic; they possess two haploid nuclei per hyphal compartment. In contrast to sporidia, the dikaryotic phase of *U. maydis* requires infection of the plant to grow and differentiate, and cannot be maintained in the laboratory.

Proliferation of the fungus inside the plant leads to disease symptoms such as chlorosis, anthocyanin formation, reduced growth, and the appearance of tumors harboring the developing teliospores.

Mature spores are released from the tumors and spread by rain and wind. Under appropriate conditions, a metabasidium is formed in which meiosis occurs. Resulting haploid nuclei migrate into elongated single cells. These cells detach from the metabasidium to become the sporidia, thus completing the life cycle.

## Host/Pathogen Conflict

Plants have evolved efficient defense systems against pathogenic microbes. A rapid plant defense reaction after pathogen attack is the oxidative burst, which involves the production of reactive oxygen species at the site of the attempted invasion. As a pathogen, *U. maydis* can respond to such an oxidative burst by an oxidative stress response, regulated by gene yap1. This response protects *U. maydis* from the host attack, and is necessary for the pathogen's virulence. Furthermore, *U. maydis* has a well-established recombinational DNA repair system. This repair system involves a homolog of Rad51 that has a very similar sequence and size to its mammalian counterparts. This system also involves a protein, Rec2 that is more distantly related to Rad51, and Brh2 protein that is a streamlined version of the mammalian Breast Cancer 2 (BRCA2) protein. When any of these proteins is inactivated, sensitivity of *U. maydis* to DNA damaging agents is increased. Also mitotic recombination becomes deficient, mutation frequency increases and meiosis fails to complete. These observations suggest that recombinational repair during mitosis and meiosis in *U. maydis* may assist the pathogen in surviving DNA damage arising from the host's oxidative defensive response to infection, as well as from other DNA damaging agents.

# Crown Gall

Crown gall is identified by overgrowths appearing as galls on roots and at the base or "crown" of woody plants such as pome (e.g., apple, pear) and stone (e.g., cherry, apricot) fruit and nut (e.g., almond, walnut) trees. Crown galls are also formed on ornamental woody crops such as roses, Marguerite daisies, and Chrysanthemum spp. as well as on vines and canes such as grapevines and raspberries. Marguerite daisies, chrysanthemums and grapevines can become infected systemically. Occasionally, galls have been observed on field crops such as cotton, sugar beets, tomatoes, beans and alfalfa, but the disease does not impact such crops economically. Crown gall is caused by Agrobacterium tumefaciens, a Gram-negative, bacilliform bacterium that is normally associated with the roots of many different plants in the field. This bacterium can survive in the free-living state in many soils with good aeration such as sandy loams where crown gall diseased plants have grown. The bacterium can also survive on the surface of roots (rhizoplane) of many orchard weeds.

Plants representing over 93 plant families are susceptible to crown gall as judged by experimental inoculations. Owing to their high susceptibility to crown gall, plants such as Jimson weed (Datura stramonium) and sunflower (Helianthus annuus) are used as assay hosts for testing the degree of virulence of A. tumefaciens. Also, Kalanchoë daigmontiana (also known as Bryophyllum) is used for assaying A. tumefaciens, but the plant is less sensitive than Datura.

## Pathogen Biology

Agrobacterium tumefaciens is a rhizoplane bacterium whose characteristics are Gram-negative, strictly aerobic, bacilliform rods measuring 1 x 3 μm, and whose nutritional requirements are non-fastidious.

The rods bear flagella that are arranged subpolarly around the cylindrical circumference of the cell, referred as circumthecal flagellation. When A. tumefaciens cells perceive plant phenolic compounds, the virulence genes that are located in the resident Ti (tumor-inducing) plasmid are expressed, resulting in the formation of a long flexuous filament called the T pilus .The activation of VirA also shuts off motility of the circumthecal flagella, presumably when A. tumefaciens cells attach to plant cells. Attachment to the plant cells is a prerequisite for initiating the transfer of the T-DNA into the plant cell. Both the circumthecal flagella and the T pilus play an essential role in virulence, presumably by bringing the bacterial cell to its target followed by attachment to the plant host, respectively.

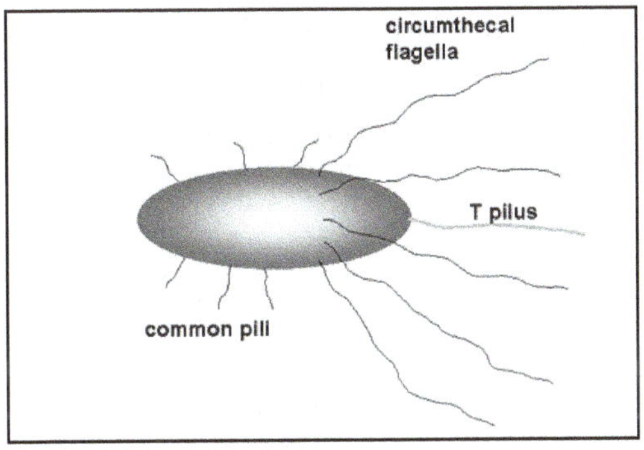


## Agrobacterium Tumefaciens Biotypes and Biovars

Based on some distinct phenotypic differences, A. tumefaciens isolates were originally classified into three biotypes or biovars (biotype I, II and III; biovar 1, 2, and 3). Biotype I or biovar 1 strains produce 3-ketosugars and usually have wide host ranges. Biotype II or biovar 2 strains mainly classify as the hairy root-forming organism, A. rhizogenes. Biotype III or biovar 3 isolates are mainly confined to grapevines, prefer L-tartaric acid over glucose and produce polygalacturonase. Because grapevine isolates formed a distinct group verified by DNA homology studies and were frequently limited in host-range to grapevines, biovar 3 strains have been reclassified into one species, A. vitis. Agrobacterium rubi strains infect canes of the genus Rubus, representing blackberry and raspberry.

## Ti Plasmid and Virulence Genes

Experimental inoculation of an assay host plant such as Jimson weed (Datura stramonium) results in tumor formation within two weeks. Virulence and the host-range of A. tumefaciens are conferred by a large extrachromosomal DNA element designated as the Ti plasmid (for tumor-inducing) that resides in all virulent strains of this pathogen. The Ti plasmid is approximately 200 kilobases in length and is comprised of a covalently closed, double-stranded DNA circular molecule.

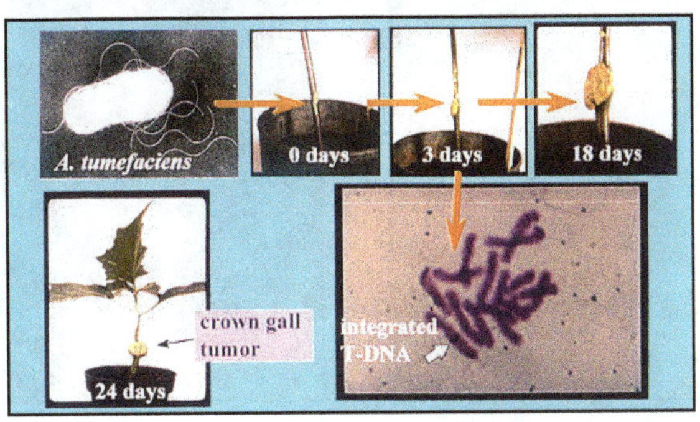

Upon recognition of plant signals in the form of dimethoxyl phenolic compounds released from plant wounds, virulence gene, virA, encodes histidine kinase (VirA) that phosphorylates the response regulator (VirG), which in turn transcriptionally activates the remaining vir genes of the vir regulon. The products of the vir genes carry out the processing of the T-DNA and functions for T-DNA transfer to the plant cell. Also, perception of the plant phenolic compounds by VirA switches off A. tumefaciens cell motility. In addition, certain chromosomal genes (e.g., chv, att genes) of A. tumefaciens are also involved in virulence. Most of them have a role in the attachment of A. tumefaciens to plant cells, in promoting signal molecule recognition, and in the regulation of certain vir and T-DNA genes.

T-DNA transfer from A. tumefaciens to plants is aided by a long flexuous appendage known as the T pilus. The T-DNA is delivered as a single-stranded DNA molecule coupled to a VirD2 protein into the plant cell. The T-DNA is integrated into the plant chromosome as evidenced by in situ hybridization using the T-DNA as the probe. The T-DNA genes are expressed by plant transcriptional machinery.

The products of the T-DNA genes catalyze the formation of auxin, cytokinin and opines. Profuse production of auxin and cytokinin in the transformed cells results in abnormal cell division, cell enlargement, and uncontrolled growth of the infected plant tissues. Opines are utilized as specific nitrogen and carbon sources by A. tumefaciens. The opines produced by the transformed plant cells provide unique nutritional sources for the pathogen, and certain opines promote transfer of the Ti plasmid between Agrobacterium strains. Certain opines such as octopine and agrocinopine serve as inducers of the Ti plasmid conjugative transfer system.

## Injury to Young Tissues Leads to Infection

Wounds made by cultivating practices include accidental, localized injuries caused by mowing and disking machines for weed removal in orchards and vineyards. The wounds are an open invitation for the initiation of the crown gall disease of the roots and the base of trees and vines. Also, tissues injured by frost are prone to infection by A. tumefaciens. In plants where A. tumefaciens resides systemically throughout the plant, such as in Marguerite daisy, grapevines and chrysanthemum, frost injury can cause a linear or confluent array of small tumors along the vascular system of the infected plant. Galls on subterranean parts originated from initial planting of diseased or infected stock. Wounds made by grafting and pruning before transplanting are also susceptible to crown gall infection.

## Crown Gall Tumors are Often Fleshy and Easily Detached

Fresh crown galls are relatively sturdy and hard. As these galls age beyond one year, they appear convoluted with cavities and become friable. Insects such as earwigs commonly reside within these cavities. It is difficult to isolate A. tumefaciens from aged galls when most of the diseased plant tissue is dead. Aged galls are easily detached. Fresh galls are the best sources of this pathogenic bacterium.

## Crown Gall Causes Stunting of Growth

Both loss of yield and stunting of growth may occur when seedlings or young cuttings are infected in the early stages of plant growth. The lack of vigor, reduction in foliage, and water stress are associated with chronically diseased root systems. When more mature tree crops become infected, secondary growths will appear from the root systems near the trunk. These "suckers" are a good sign that the root system is infected.

## Disease Cycle and Epidemiology

## Disease Cycle and Epidemiology

Agrobacterium tumefaciens naturally resides on the rhizoplane of woody and herbaceous weeds. Its presence in soils originates from galls that were broken or sloughed off from infected plants during cultivation practices or disseminated as infected plant material. Irrigation aids in further dissemination of the A. tumefaciens bacterial cells. Agrobacterium tumefaciens is also spread by infected and infested planting stocks originating as nursery stock from uncertified sources. Secondary spread then occurs through pruning and cultivation equipment, particularly when galls are removed manually with the same cutting tools used in pruning. Tilling equipment can be contaminated by cutting through galls at or near the base of trunks of infected trees. Rogueing (removal) of infected trees and replanting in the same spot where the infected tree had grown is poor practice because sloughed off galls serve as sources of abundant populations of A. tumefaciens.

Although a single cell can be transformed and grow into a gall, the size of the initial tumor depends on the amount of inoculum. Thus, if a large number of cells of A. tumefaciens enter a plant wound, disease will be more severe and a large gall will be seen rapidly growing at the infection site. Plants that are systemically infected (such as grapevines) will harbor A. tumefaciens for extended periods of time in the absence of overt symptoms. Tissue injuries induced by accidental wounds made during cultivation practices or by frost will elicit new infections with the appearance of numerous galls along the vascular system of the host.

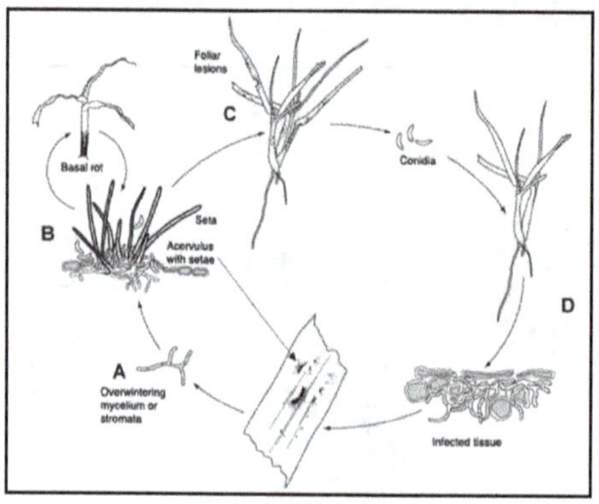


## Crown Gall Management

## Preplanting Management Options

Planting stocks: Visual examination for crown gall tumors has been the conventional primary screen for diseased material. The method is limited for complete disease control because A. tumefaciens can reside on the rhizoplane and systemically in certain host plants such as grapevines, chrysanthemums, and marguerite daisy.

Site selection: Fields that have grown cereal crops for a long period are favored as crown gall-free sites. Fields previously used for growing fruit and nut crops can remain infested with A. tumefaciens. Certain weeds such as morning glory (Ipomoea leptphylla) can serve as natural hosts of A. tumefaciens and therefore perpetuate the survival of this pathogen in field soils.

Crop rotation: A crop rotation program employing cereal crops followed by green manuring helps reduce the population size of A. tumefaciens.

Chemical eradicants: Eradication of crown gall using creosote-based compounds, copper-based solutions, and strong oxidants such as sodium hypochlorite are transiently effective. The chemical eradicant application procedure is labor intensive and therefore costly both monetarily and to the environment. The superficial treatments are ineffective against systemically infected plants. Generally, chemicals are rarely used for control of crown gall.

Biological control: Certain strains of A. tumefaciens are sensitive to the antibiotic agrocin produced by A. radiobacter, a closely related soil-borne bacterium that does not infect plants. An example of an antibiotic produced is Agrocin-84, which is an analog of the opine agrocinopine A. Agrocinopine A is produced in crown gall tumors induced by A. tumefaciens strains whose Ti plasmid encodes for nopaline and agrocinopine A. Agrocin-84 mimics agrocinopine A and therefore is taken up by the same transport system used by A. tumefaciens to utilize agrocinopine A. Inside the A. tumefaciens cell, the antibiotic Agrocin-84 inhibits DNA replication and cellular growth.

Plants are protected against Agrocin-84 sensitive strains of the pathogen by dipping the root system into a suspension of A. radiobacter K84 before planting in infested fields. Biological control of crown gall has been a very effective method to control crown gall in several locations. In many other regions where agrocin-insensitive strains of A. tumefaciens (strains that do not acquire agrocinopine A) reside, this biological control strategy is ineffective.

Genetically engineering: Transgenic crop plants harboring one or more unique genes tailored to protect the plant from crown gall have been developed. Genes encoding products that degrade or inactivate the T-DNA strand complex when it enters the host cell, that prevent the expression of T-DNA genes encoding indoleacetic acid and cytokinin biosynthesis, and that prevent A. tumefaciens attachment to its target are some examples currently being tested. Biotechnology companies, such as DNA Plant Technology.

(Oakland, CA), are applying sense strand messenger RNA or small-interfering RNAs to develop crown gall resistant fruit and nut crops.

### Management in Established Fields

Trees in fruit and nut orchards can be maintained over long periods if the trees became infected at maturity. Diseased trees will bear crop, but with age the trees will become unthrifty and suffer dehydration as their root system becomes progressively infected. The removal of infected trees and vines is costly in loss in time and in money. Annual row crops such as sugar beets and cotton occasionally will have a few plants with crown gall, but the disease is considered of low economic importance. Perennial field crops such as alfalfa will occasionally become infected with A. tumefaciens where the organism is spread by the mowing equipment. Usually, rogueing of the diseased plants is sufficient to minimize further spread of crown gall.

## Damping Off

Damping-off or damping off is a destructive disease of plant seedlings. Damping-off is caused by a number of seed- and soil-borne fungi and funguslike oomycetes, including Rhizoctonia solani, Aphanomyces cochlioides, and species of Pythium, Phytophthora, Botrytis, Fusarium, Cylindrocladium, Diplodia, Phoma, and Alternaria. Given the diversity of the pathogens, damping-off can affect a wide range of plant species and can cause losses for a number of economically important food crops.

There are two types of damping-off: preemergence, in which sprouting seeds decay in soil and young seedlings rot before emergence; and postemergence, in which newly emerged seedlings suddenly wilt, collapse, and die from a soft rot at the soil line. Woody seedlings wilt and wither but remain upright; root decay often follows. Greatest losses occur in cold wet soils in which germination and emergence are slow, often in indoor conditions.

Damping-off can be avoided by starting seed in light, well-drained, well-prepared soil or sterile mix (containing perlite, peat moss, or vermiculite); treating soil with steam, dry heat or a fumigant; avoiding overcrowding, excessive shade, overwatering, too deep planting, and overfertilizing; and sowing crack-free, healthy seed dusted with fungicide seed protectant. An early outbreak can be controlled by applying a fungicide solution.

## Causes and Prevention

Plant pathogens exist everywhere in the environment, but they thrive under certain conditions caused by:

- Reusing dirty containers
- Sowing a seed too deep
- Overwatering
- Overcrowding
- Wet leaves
- Too much humidity
- Other environmental stress, like low light or cool conditions

## To Minimize the Chances of Damping Off

- Start with clean pots, plant markers, and tools that have been washed with hot water and soap.
- Use clean potting soil that was not infested with disease last season.
- Make sure your potting soil is light and fast-draining.
- Sow your seeds on top of the soil, then add a layer of vermiculite or very fine granite (like chick grit) to cover; the drier surface is less likely to harbor fungi.
- Thin your seedlings as needed and provide good air circulation around them.
- Water your pots from the bottom up to avoid wetting the stems and leaves.
- Keep the soil evenly moist but not water-logged.

Since the key to treating damping off is preventing the infection in the first place, it might make sense to preempt the problem by applying chamomile tea, clove tea, or a sprinkle of cinnamon to your soil. (All of these remedies are known for their antifungal properties.)

## Downy Mildew

Downy mildew is a disease of plants, especially in cool humid regions, caused by several funguslike organisms of the phylum Oomycota. White, gray, bluish, or violet downy patches of mildew form mostly on the undersides of leaves in damp weather. Pale green to yellow or brown areas usually develop on the upper leaf surface opposite the downy growth. Affected leaves often wilt, wither, and die early. Stems, flowers, and fruits are sometimes infected. Seedlings may wilt and collapse. Garden plants, bush fruits, vegetables, and certain trees, shrubs, field crops, and weeds are susceptible. Downy mildew is commonly caused by members of the oomycete genus Sclerospora, but other pathogens include species of Bremia, Peronospora, Phytophthora, Plasmopara, and Pseudoperonospora.

Downy mildew can be avoided by rotating annual flowers and vegetables and by avoiding overwatering, overcrowding, and poorly drained soil. Other avoidance measures are growing resistant varieties, sowing disease-free seed, removing diseased parts and crop refuse, eliminating weeds, and maintaining balanced soil fertility. The application of copper or either of the fungicides maneb or zineb is effective against many downy mildews, but the amount of residue on vegetables must be considered.

### Symptoms

Downy mildew colonies often appear first on the underside of leaves, and they sometimes have a bluish tinge. In many cases, they can grow systemically throughout the plant. If growing abundantly on a leaf, downy mildew colonies can be confused with gray mold (Botrytis) or with powdery mildew. Microscopically, they are very easy to tell apart from powdery mildew and Botrytis. On the foliage, small yellow spots develop on the upper sides of the leaf while white to bluish-white fluffy growth forms on the underside of the leaf. As the leaf spot dies, the fluffy growth darkens to gray in color. Infected leaves and branches may be distorted and die.

### Treatment and Management

### Cultural Options

Because the downy mildew pathogen does not overwinter in midwestern fields, crop rotations and tillage practices do not affect disease development. The pathogen tends to become established in late summer. Therefore, planting early season varieties may further reduce the already minor threat posed by downy mildew.

### Chemical Control

Fungicides applied specifically for downy mildew control may be unnecessary. Broad spectrum protectant fungicides such as chlorothalonil, mancozeb, and fixed copper are at least somewhat effective in protecting against downy mildew infection. Systemic fungicides are labeled for use against cucurbit downy mildew, but are recommended only after diagnosis of this disease has been confirmed. In the United States, the Environmental Protection Agency has approved oxathiapiprolin for use against downy mildew. In Canada, a mixture of zoxamide and mancozeb was registered for control of the mildew under the trademark Gavel (fungicide).

## Organic Control

One way to control downy mildew is to eliminate moisture and humidity around the impacted plants. Watering from below, such as with a drip system, and improve air circulation through selective pruning. In enclosed environments, like in the house or in a greenhouse, reducing the humidity will help as well.

## Plant Specific Mildews

### Basil

Downy mildew of basil caused by *Peronospora belbahrii* has been a huge problem for both commercial producers and home growers. The disease was first reported in Italy in 2004, was reported in the U.S. in 2007 and 2008 and has been steadily increasing in prevalence, distribution, and economic importance since then.

### Cucurbitaceae

Cucurbitaceae downy mildew (caused by *Pseudoperonospora cubensis*) is specific to cucurbits (e.g., cantaloupe (*Cucumis melo*), cucumber (*Cucumis sativus*), pumpkin, squash, watermelon (*Citrullus lanatus*) and other members of the gourd family). The disease is one of the most significant diseases of cucurbits worldwide.

### Grapes

*Plasmopara viticola* is the causal agent of grapevine downy mildew.

### Hops

Hop Downy Mildew (caused by *Pseudoperonospora humuli*) is specific to hops (*Humulus lupulus*). The disease is the single most devastating disease in Western United States hopyards, since the microbe thrives in moist climates. Infected young hop bines become stunted with thickened clusters of pale curled leaves. These spikes have a silvery upper surface, while the undersides of leaves become blackened with spores. These dwarfed spikes are called "basal spikes". 'Lateral' or 'terminal' spikes occur further up the vine. An entire hop crop could be devastated in only a few days.

### Ornamentals

A new and particularly aggressive form of impatiens downy mildew has recently emerged as a major threat to the cultivation of ornamental impatiens in the United States, where they are one of the most popular ornamental plants.

### Soybeans

*Peronospora manshurica* infects soybeans, reducing photosynthetic activity, yield, and quality. The fungus spreads by oospores on diseased leaves and on infected seed. The disease spreads in environments with high humidity and favors temperatures between 20-22° C. Tufts of grayish to pale-colored sporangiophores on the underside of leaves easily distinguish the infection from other foliar diseases. The disease is often controlled using the fungicides mancozeb, maneb, or zineb.

### Sunflowers

*Plasmopara halstedii* infects sunflowers, producing oospores which can remain dormant in the soil for many years.

## Early Blight

Early blight has been known to target tomatoes, affecting the fruits and stems alike. In most cases, this ailment tends to affect older plants but this disease has also been known to infect seedlings.

### Symptoms

Symptoms of early blight occur on fruit, stem and foliage of tomatoes and stem, foliage and tubers of potatoes. Initial symptoms on leaves appear as small 1-2 mm black or brown lesions and under conducive environmental conditions the lesions will enlarge and are often surrounded by a yellow halo. Lesions greater than 10 mm in diameter often have dark pigmented concentric rings. This so-called "bullseye" type lesion is highly characteristic of early blight. As lesions expand and new lesions develop entire leaves may turn chlorotic and dehisce, leading to significant defoliation. Lesions occurring on stems are often sunken and lens-shaped with a light center, and have the typical concentric rings. On young tomato seedlings lesions may completely girdle the stem, a phase of the disease known as "collar rot," which may lead to reduced plant vigor or death.

Infection of both green and ripe tomato fruit normally occurs through the calyx with lesions sometimes reaching a considerable size. The lesions appear leathery and may have the characteristic concentric rings. Infected fruit will frequently drop prematurely. Symptoms on potato tubers are characterized by sunken, irregular lesions, which are often surrounded by a raised purple border. Beneath the surface of the lesion the tuber tissue is leathery or corky with a brown discoloration. Early blight lesions on tubers tend to be dry and are less prone to invasion by secondary organisms than lesions of other tuber rots. After prolonged storage severely diseased tubers may become shriveled.

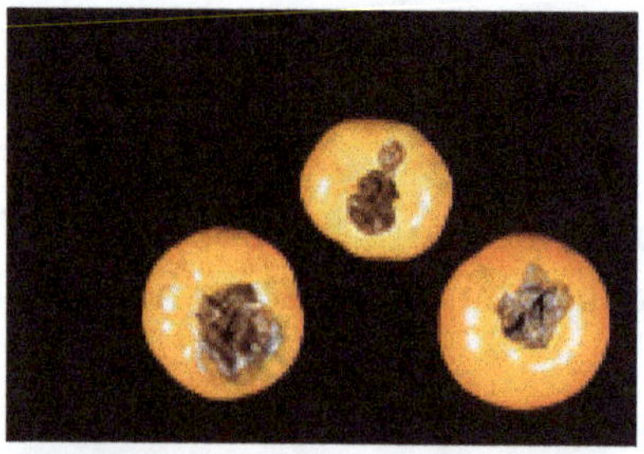

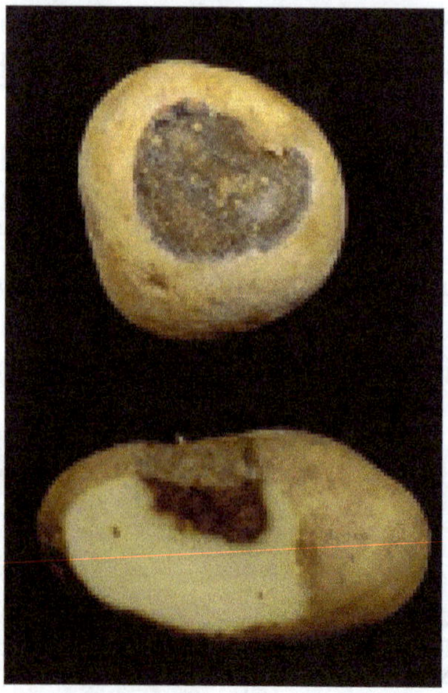

## Pathogen Biology

The causal pathogen of early blight is the fungus Alternaria solani. There is no known sexual stage and hence it is classified as a Deuteromycete. The genus Alternaria is a large and important group of pathogenic fungi, which cause a significant number of important diseases. The fungus is readily cultured on artificial media such as V8 juice where it produces a deeply pigmented gray/black hairy colony. The mycelium is haploid and septate, becoming darkly pigmented with age. Sporulation in culture can be stimulated by exposure to fluorescent light. The asexual conidia are borne singly or in a chain of two on distinct conidiophores. The beaked conidia normally possess 9–11 transverse septae. Morphological and pathogenic variability among isolates of A. solani has given rise to claims of the existence of races, although this remains unproven.

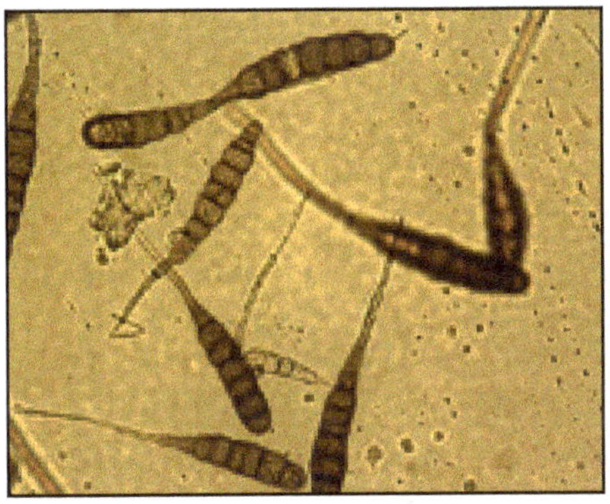

## Disease Cycle and Epidemiology

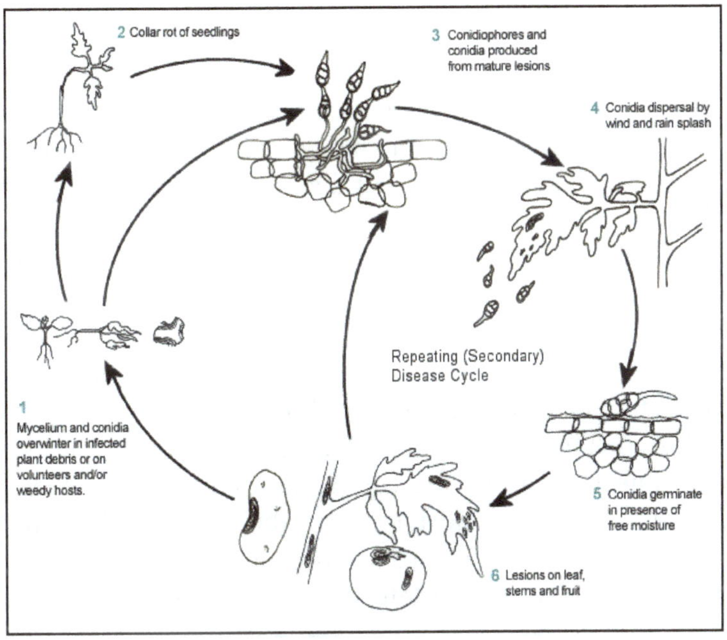


Alternaria solani overwinters primarily on infected crop debris. The dark pigmentation of the mycelium increases resistance to lysis which extends the survival time in the soil to several years. Thick-walled chlamydospores have been reported, but they are found infrequently. In mild climates the pathogen can survive from season to season on volunteer tomato and potato plants as well as other weedy Solanaceous hosts such as horsenettle and nightshade.

Warm, humid (24-29° C/75-84° F) environmental conditions are conducive to infection. In the presence of free moisture and at an optimum of 28-30° C (82-86° F), conidia will germinate in approximately 40 min. Desiccated germ tubes are able to renew growth when re-wetted, and, hence, infection can occur under conditions of alternating wet and dry periods. Germ tubes penetrate the leaf epidermis directly or enter through stomata. Infection of potato tubers usually occurs through wounds in the tuber skin inflicted during harvest. Wet conditions at harvest provide a favorable environment for spore germination as well as causing swollen lenticels on the tubers which are easily invaded.

Time from initial infection to appearance of foliar symptoms is dependent on environmental conditions, leaf age, and cultivar susceptibility. Early blight is principally a disease of aging plant tissue. Lesions generally appear quickly under warm, moist conditions on older foliage and are usually visible within 5-7 days after infection.

A long wet period is required for sporulation but it can also occur under conditions of alternating wet and dry periods. Conidiophores are produced during wet nights and the following day light and dryness induce them to produce spores, which emerge on the second wet night.

Secondary spread of the disease results from conidia being dispersed mainly be wind and occasionally by splashing rain or overhead irrigation. Early blight is considered polycyclic with repeating cycles of new infection. This is the period when the disease has the potential to spread rapidly and build up to damaging levels in the crop.

## Disease Management

### Cultural Practices

In many cases employing sound cultural practices that maintain potato and tomato plants in good health will keep early blight losses below economic levels. Because the pathogen over winters on infected crop debris, in-field sanitation procedures that reduce initial inoculum in subsequent crops are beneficial. Consideration should be given to removing potentially infected material such as decaying vines and fruits from the vicinity of production fields. Controlling volunteers and weeds, such as nightshade and horse nettle which serve as alternative hosts for the disease, prior to planting the new crop will help to reduce the risk of transmission of disease. Ensuring seed or transplants are pathogen free before placing out in the field and rotating fields to a non-susceptible host crop will also help to reduce buildup of inoculum in the soil. Optimal tuber maturity is the most important factor for control of tuber infection. Tubers harvested before maturing are susceptible to wounding and infection. Tuber infection can be reduced by careful handling during harvest to minimize wounding as well as avoiding harvesting during wet conditions if possible. Tubers should be stored at 50 to 55 F, at high relative humidity and with plenty of aeration to promote wound healing which will reduce the amount and severity of tuber infections that develop in storage.

During the season, overhead irrigation schedules should minimize the duration of leaf wetness in the crop. Avoid irrigation in cool, cloudy periods or late in evening when foliage may stay wet for extended periods. Selecting fields with good drainage and an absence of natural impediments to air flow over the crop, e.g. rows of trees, will reduce periods of leaf wetness. Maintenance of adequate soil fertility levels is also critical for managing early blight. The disease is often associated with crops suffering from a lack of nitrogen, particularly towards the end of the growing season on older senescing foliage. Management of other diseases such as Verticillium wilt will reduce plant stress and hence, early blight severity.

### Resistant Cultivars

Complete resistance to early blight does not exist in commercial potato or tomato cultivars. Using wild Lycopersicon species which show a high degree of resistance in breeding programs has led to the release of a number of cultivars of potato and tomato with a degree of resistance to early blight. Apparent levels of resistance are often correlated with plant age. Immature potato and tomato plants are relatively resistant to early blight but, after tuber and fruit initiation, susceptibility increases gradually, and mature plants are very susceptible to the pathogen.

### Chemical Control

Fungicides with protectant and curative properties are registered for use against early blight on tomato and potato. The cheaper protectant fungicides such as mancozeb and chlorothalonil are the foundation of most early blight management programs. These fungicides must be reapplied every 7-10 days to provide protection of new growth as well as to counter the effects of weathering which progressively removes the chemical from the leaf surface. Advantages of these types of products include their reliable efficacy and multi-site mode of action, which reduces the risk of resistant isolates developing in the pathogen population; therefore, they are useful as tank mix partners or used in rotation with other fungicides. Disadvantages include the need to apply regularly and their relatively high use rates.

## Fire Blight

Named for the scorched appearance of infected leaves, fire blight is a destructive bacterial disease (Erwinia amylovora) found on apples, pears and other members of the rose family. The disease enters the tree at the tips of the branches and then travels down the stems causing dieback. It attacks soft new growth first, so you would notice dieback at the top of the plant.

Most infected leaves and branch tips wilt rapidly turn brown or black; the leaves die but do not drop off. Trees will also develop reddish water soaked lesions on the bark. On warm days, these lesions ooze an orange-brown liquid. Fire blight kills blossoms, shoots, limbs and sometimes, the entire tree.

The bacteria overwinters in infected bark and is spread by splashing rain, dew, wind and insects. It spreads rapidly in moist, warm weather, especially during bloom.

## Symptoms

Fire blight on a pear tree.

Tissues affected by the symptoms of *Erwinia amylovora* include blossoms, fruits, shoots, and branches of apple (Pomoideae), pear, and many other rosaceous plants. All symptoms are above ground and are typically easy to recognize. Symptoms on blossoms include water soaking of the floral receptacle, ovary, and peduncles. This results in a dull, gray-green appearance at 1–2 weeks after petal fall, and eventually tissues will shrivel and turn black. The base of the blossom and young fruit show similar symptoms as infection spreads. Opaque white or amber colored droplets of bacterial ooze can be seen on the infected tissue when the environment is high in humidity. Shoots show similar symptoms but develop much more rapidly. A "Shepherd's Crook," can be seen when the tip of the shoot wilts, diseased shoot leaves typically have blackening along the mid-vein and then die. In number, diseased shoots give the tree a blighted appearance. Initial infection of blossoms and shoots can spread to larger tree limbs. Branches will darken and become water soaked. Advanced infection develops cracks in bark and a sunken surface. Wood under the bark will become streaked with black discoloration. Immature fruit forms water-soaked lesions and later turned black. Bacterial ooze can be found on these lesions. Severe infections result in fruit turning entirely black and shriveling. A primary inoculum of this disease is typically from cankers formed the season before. The factors that determine whether or not cankers become active are not well known, but it is thought that cankers found on larger tree limbs are more likely to become active. It is also thought that age may be a factor.

## Dissemination

Gala apple branch with "scorched" leaves after a severe fire blight infection.

Honeybees and other insects, birds, rain and wind can transmit the bacterium to susceptible tissue. Injured tissue is also highly susceptible to infection, including punctures and tears caused by plant-sucking or biting insects. Hailstorms can infect an entire orchard in a few minutes, and growers do not wait until symptoms appear, normally beginning control measures within a few hours.

Once deposited, the bacterium enters the plant through open stomata and causes blackened, necrotic lesions, which may also produce a viscous exudate. This bacteria-laden exudate can be distributed to other parts of the same plant or to susceptible areas of different plants by rain, birds or insects, causing secondary infections. The disease spreads most quickly during hot, wet weather and is dormant in the winter when temperatures drop. Infected plant tissue contains viable bacteria, however, and will resume production of exudate upon the return of warm weather in the following spring. This exudate is then the source for new rounds of primary infections.

The pathogen spreads through the tree from the point of infection via the plant's vascular system, eventually reaching the roots and graft junction of the plant. Once the plant's roots are affected, the death of the plant often results. Over pruning and over fertilization (especially with nitrogen) can lead to watersprout and other midsummer growth that leave the tree more susceptible.

*E. amylovora* typically makes its entry into its host xylem or cortical parenchyma. It can also enter through stomata, lenticles and hydathodes. It is dispersed by rain and or insects naturally, but this mode of dispersal is very ineffective and can only be effective for local transmission of the pathogen. Aerosols are also suspected in playing a role in its transmission due to the detection of *E. amylovora* in Mediterranean regions. In composition the pathogen is composed of short rods with rounded ends made motile by many peritrichous flagellae. *E. amylovara* is a gram negative bacterium.

## Management

Spraying plants with streptomycin or injecting plants with oxytetracycline can prevent new infections. The widespread use of streptomycin spray has led to antibiotic resistance in some areas, such as California and Washington. Certain biological controls consisting of beneficial bacteria or yeast can also prevent fire blight from infecting new trees. The only effective treatment for plants already infected is to prune off the affected branches and remove them from the area. Plants or trees should be inspected routinely for the appearance of new infections. The rest of the plant can be saved if the blighted wood is removed before the infection spreads to the roots. There is no known cure; prevention is the key.

*E. amylovora* needs to be destroyed externally, before it enters the cell. This is simply because once it enters the host, it spreads during the endophytic phase of pathogenesis. Once this happens external control methods become ineffective. The ideal control method is to apply copper and antibiotics to the plant externally. This is the only effective method and it is indeed preventative. Currently it has been noted that *E. amylovora* has developed a resistance to the antibiotic streptomycin, as do most bacteria due to their flexible ability to transfer preferential genes promoting resistance to certain antibiotics horizontally from species not even similar to it as all bacteria can.

Phytosanitary measures have been employed as the best sanitary measures against *E. amylovora* dispersal. High risk countries are encouraged not to import plants susceptible to the pathogen into their territory because, once the bacteria become established in an area it is nearly impossible to eradicate the disease. Nurseries and orchards in such regions are placed on strict phytosanitary surveillance measures and well-monitored. Imported and infected crops are destroyed as soon as they are noticed since the bacteria spreads very rapidly and eradication methods are usually costly and inefficient.

### Importance

Besides the historical importance of being the first bacterium proven to be a plant pathogen, it is extremely economically important. Control and loss costs are estimated approximately $100 million a year in the USA. Specifically, in Michigan in the year 2000, $42 million in losses is estimated because of the removal of about 400,000 apple trees. Warm, humid, and wet weather in May resulted in this epidemic. While approximately $68 million is estimated in losses in Washington and northern Oregon. *E. amylovora* is spread all through the USA and worldwide causing severe damage although it is unlikely to cause severe damage in northern Europe. As long as *E. amylovora* is not introduced to Central Asia where wild apple trees still grow, it will not modify any ecosystems. Biodiversity is not impacted either, as no plant species are threatened with extinction due to this pathogen. Growing pears in Emilia-Romagna in Italy is a traditional activity for some families, and fire blight threatens this tradition that has been passed down for several generations. In southern Germany apple and pear trees have been a part of the landscape for a long time, and are difficult to protect. The decline of apple and pear trees from their landscape can be expensive to replace and could have a negative effect on tourism. In the long-run, fire blight is a very important factor of economy and society.

### Pathogenesis

Pathogenicity depends on many different factors such as the production of the siderophore desferrioxamine, metalloproteases, plasmids, and histone-like proteins. However, some essential factors of pathogenicity are variations in the synthesis of extracellular polysaccharides (EPS) and the mechanism of type III secretion system and its associated proteins. EPS helps bacterial pathogens avoid plant defenses, "clog" the host's vascular system, protect bacteria against desiccation and attach to both surfaces and one another. One EPS is amylovoran, a polymer of pentasaccharide repeating units. If a strain of *E. amylovora* can not produce amylovoran it is not pathogenic and can not spread in plants. Levan is another EPS, and a lack of it will slow development of symptoms. Type III secretion systems are used for exporting and delivering effector proteins into the cytosol of host plants. This system mainly consists of Hrc proteins. Motility is another major virulence factor. Since *E. amylovora* is not an obligate biotroph, it is able to survive outside the host which allows it to spread in many ways such as rain.

## Leaf Curl

Peach leaf curl, also known as leaf curl, is a disease caused by the fungus Taphrina deformans. Peach leaf curl affects the blossoms, fruit, leaves, and shoots of peaches, ornamental flowering peaches, and nectarines, and is one of the most common disease problems for backyard gardeners growing these trees. The distorted, reddened foliage that it causes is easily seen in spring. When severe, the disease can reduce fruit production substantially.

Peach leaf curl symptoms can cause leaves to curl and distort.

Though uncommon, symptoms on fruit can occur, making the surface corky and cracked.

Peach leaf curl symptoms typical of a serious infection.

## Identification and Damage

Peach leaf curl first appears in spring as reddish areas on developing leaves. These areas become thickened and puckered, causing leaves to curl and severely distort. The thickened areas turn yellowish and then grayish white, as velvety spores are produced on the surface by the leaf curl fungus. Later affected leaves turn yellow or brown and can remain on the tree or may fall off; they are replaced by a second set of leaves that develop more normally unless wet weather continues. The loss of leaves and the production of a second set result in decreased tree growth and fruit production. Defoliation in spring may expose branches to sunburn injury.

The peach leaf curl pathogen also infects young green twigs and shoots. Affected shoots become thickened, stunted, distorted, and often die. Only rarely do reddish, wrinkled to distorted (or hypertrophied) areas develop on fruit surfaces. Later in the season these infected areas of fruit become corky and tend to crack. If leaf curl infection builds up and is left uncontrolled for several years, the tree may decline and need to be removed.

## Life Cycle

Leaf symptoms appear about 2 weeks after leaves emerge from buds. The fungus grows between leaf cells and stimulates them to divide and grow larger than normal, causing swelling and distortion of the leaf. Red plant pigments accumulate in the distorted cells. Cells of the fungus break through the cuticle of distorted leaves and produce elongated, sac-like structures called asci that produce sexual spores called ascospores, which give the leaf a grayish white, powdery or velvet like appearance. The ascospores are released into the air, carried to new tissues, and bud (divide) to form bud-conidia.

The fungus survives the hot, dry summer as ascospores and bud-conidia (asexual spores) on the tree's surfaces. When the weather turns cool and wet in fall, the ascospores germinate to produce more bud-conidia. The new and old bud-conidia continue to increase in number by budding. Eventually a film of bud-conidia is formed on the tree's surface. In spring, the bud-conidia move by splashing water from irrigation or rain and can infect new leaves.

Periods of cool, wet weather, when leaves are first opening on the tree, favor the disease. The optimum temperature for fungal growth in laboratory cultures is 68° F, the minimum is 48° F, and the maximum is 79° to 87° F. Budding of bud-conidia occurs at or above 95% relative humidity. Wetness from rain, dew, or irrigation for more than 12.5 hours at temperatures below 61° F is needed for infection. Maximum infection occurs when trees are wet for 2 or more days. Although leaves can be infected, symptoms might not appear if temperatures remain above 69° F. Cool weather prolongs the period of disease development by favoring the pathogen and slowing leaf growth. Development of peach leaf curl ceases when young tissue is no longer developing or when weather turns dry and warmer (79° to 87° F).

## Management

To prevent peach leaf curl, use resistant peach and nectarine varieties where possible. For nonresistant varieties, treat trees with a fungicide every year after leaves have fallen. In cooler northern locations leaf fall usually is in late November. In warmer southern locations leaf fall can be as late as early January. Generally a single early treatment when the tree is dormant is effective, although in areas of high rainfall or during a particularly wet winter, it might be advisable to apply a second spray late in the dormant season, preferably as flower buds begin to swell but before green leaf tips are first visible.

## Resistant Varieties

A few peach varieties are available that are resistant or partially resistant to leaf curl. Currently available resistant varieties include Frost, Indian Free, Muir, and Q-1-8.

The peach cultivar Frost is reportedly very tolerant but must receive fungicide applications the first 2 to 3 years. Redhaven peach and most cultivars derived from it are tolerant to peach leaf curl, whereas Redskin peach and cultivars derived from it range from susceptible to highly susceptible to the disease.

There are fewer resistant nectarines, although Kreibich is one such variety.

## Fungicides

### Copper

Historically, the most commonly used fungicides available to home gardeners have been the fixed copper products. For all copper-containing products, the active ingredient, copper, is listed as "metallic copper equivalent," or MCE, on the label. Various product formulations differ widely in their metallic copper content. The higher the MCE, the greater the amount of copper and the more effective the product will be. However, other factors such as coverage, use of additives as such stickers and spreaders, and frequency and duration of rain, which can wash off the copper, also will impact product effectiveness. In all cases, the copper is active only when it is wet, when the copper ions are in solution.

Fixed copper products include tribasic or basic copper sulfate, cupric hydroxide, and copper oxychloride sulfate (C-O-C-S), but currently only liquid products containing copper ammonium complex products with 8% MCE (e.g., Kop R Spray Concentrate and Liqui-Cop) are available to consumers. The most effective copper product, 90% tribasic copper sulfate with a 50% MCE (Microcop), is no longer available to retail outlets, because the manufacturer withdrew the product in 2010, although remaining supplies still can be sold.

The copper ammonium complex products can be made more effective by adding 1% horticultural spray oil to the application mix; the oil also aids in controlling some aphids, scale insects, and mites. Copper soap (copper octanoate) fungicides are also available, and preliminary research indicates they may provide some protection of trees.

Be aware that repeated annual use of copper products over many seasons can result in a buildup of copper in the soil, which eventually can become toxic to soil organisms, and if it moves into waterways, can harm some aquatic species.

### Bordeaux Mixture

Copper sulfate is not a fixed copper and, when used alone, is less effective than tribasic copper sulfate or other fixed copper products. However, if copper sulfate is mixed with hydrated lime to make a Bordeaux mixture, the copper sulfate and calcium in the lime react together to form a fixed copper product that is effective against peach leaf curl. Bordeaux mixture is not available for sale; it must be mixed up just before application, and the ingredients can be very difficult to find. For information on preparing Bordeaux mixture.

### Other Fungicides

The synthetic fungicide chlorothalonil currently is the only other noncopper fungicide available for managing peach leaf curl on backyard trees. Lime sulfur (calcium polysulfide) products no longer are registered for backyard use.

### Application

Thorough coverage with any fungicide is essential to obtain adequate disease control. Trees should be sprayed to the point of runoff or until they are dripping.

When using pesticides, always read and follow the label for usage, rates, toxicity, and proper disposal. Proper protective clothing and gear including goggles should be used when handling any pesticides.

### Physical Controls

Although symptoms of leaf curl are seen primarily in spring as new leaves develop, there is little you can do to control the disease at this time. Some people remove diseased leaves or prune infected shoots, but this has not been shown to improve control. Normally, diseased leaves fall off within a few weeks and are replaced by new, healthy leaves, unless it is rainy.

If a tree is severely affected with peach leaf curl this can stunt its growth, so consider thinning fruit later in the season. Pruning in fall prior to applying any fungicides can reduce spore numbers overwintering on the tree and reduce the amount of fungicide needed. If leaf curl symptoms occurred on your trees in spring, be sure to treat the following fall and winter to prevent more serious losses the following year.

## Leaf Spot

Leaf spot is a common descriptive term applied to a number of diseases affecting the foliage of ornamentals and shade trees. The majority of leaf spots are caused by fungi, but some are caused by bacteria. Some insects also cause damage that appears like a leaf spot disease. Leaf spots on trees are very common and generally do not require spraying. Leaf spot may result in some defoliation of a plant. An established plant can tolerate almost complete defoliation if it happens late in the season or not every year. Small or newly planted trees that become defoliated are more at risk of suffering damage until they become established.

### Causes of Leaf Spot

- The majority of leaf spot diseases are caused by fungi, but a few diseases are caused by bacteria or other pathogens.
- Many pathogens are somewhat host-specific and will only cause disease on trees in the same family.
- Almost all trees and shrubs are susceptible to one or more leaf spot diseases.

## Identification

- Leaf spots come in a wide variety of shapes, sizes and colors.
- Leaf spot diseases usually are seen first on the lower and inner branches where humidity is higher and leaves are shaded.
- Spots occur randomly on the leaf surface because the pathogens that cause leaf spots are blown there by wind or splashed by rain or irrigation.
- Depending upon the pathogen, leaf spots may occur on the upper, lower or both surfaces of the leaves.
- Leaf spots may be angular or rounded, raised or sunken, and have smooth or fringed edges.
- Colors can range from yellow to yellow-green to orange-red to light tan, brown or black.
- A variety of sizes of leaf spots may be observed on one plant.
- Smaller leaf spots are younger infections. Larger leaf spots are older infections.
- At the center of larger leaf spots, it is possible to see signs of the pathogen such as fungal spores or spore-producing structures.
- Some leaf spot pathogens cause leaves to drop prematurely, resulting in the tree or shrub losing most or all of its leaves.

## Life Cycle of Leaf Spot Pathogens

- The pathogens that cause leaf spot diseases can survive winter in infected leaf debris.
- Some also infect buds and young twigs.
- During the growing season, wind and splashing rain carry spores of the pathogen to susceptible plant tissue and spread the disease throughout the canopy.
- Most leaf spot diseases need either water on the leaves or very high humidity for a prolonged period of time (12 to 24 hours) to start an infection.
- Leaf spots often mature in one to two weeks.
- Mature leaf spots produce spores or bacteria that can be spread throughout the canopy, which can start a second set of leaf spots or cause new infections on other plants.
- This cycle of infection and spore production repeats whenever weather conditions are favorable.
- In years with very high humidity or frequent rain events, leaf spot pathogens can spread throughout a tree or shrub's canopy resulting in severe disease.

## Leaf Spots and Shoot Blight or Canker

Many leaf spot pathogens are only able to produce symptoms in leaf tissue. But some leaf spot pathogens can also cause blight or cankers of twigs.

Venturia shoot blight on poplar.

## Blight

Blight is a progressive dieback of young, green shoots. Leaf spot pathogens that cause dieback of young shoots typically do not progress to infect the older woody branches.

Examples of leaf spot diseases that progress to leaf and shoot blight include Aschochyta blight on lilac and Venturia shoot blight on Populus species such as poplar, aspen and cottonwood trees.

## Canker

Septoria leaf spot and canker on poplar.

Cankers are infections of shoots and branches that start as round to oval discolored areas where the bark has been killed.

- As the infection progresses cankers girdle the stem resulting in wilt and death of any leaves above the canker.
- Branches infected with canker causing fungi often have discolored bark that may be cracked or oozing gums or dark-colored sap.
- Spore-producing structures often emerge along the margins or on the face of the cankered area.

Examples of diseases that result in both leaf spots and cankers in Minnesota include Septoria leaf spot and canker of poplar and Septoria leaf spot and canker of dogwood.

## Management

- Trees and shrubs should be carefully scouted for discolored areas of bark.
- Any infected branches or shoots should be pruned out several inches below visible signs of the infection.
- Infected branches should be burned or buried.

## Bacterial Leaf spot

## Symptoms of Bacterial Leaf Spot

Bacterial leaf blight of lilac

- Bacterial leaf spot diseases often start as small dark brown to black spots with a halo of yellow tissue surrounding each spot.
- In some cases, the center of the leaf spot will dry up and fall out, giving the leaf a "shot hole" appearance.
- If weather conditions remain favorable for disease, some bacterial leaf spots will grow together creating large black blotches on leaves or turning leaves completely black.
- Shoots, buds and flowers can also become black and blighted by bacterial leaf spot pathogens.
- Bacterial blight of lilac is a common bacterial disease in Minnesota.

## Causes

- Bacterial leaf spot diseases are most commonly caused by Pseudomonas spp. or Xanthomonas spp.

- Bacterial plant pathogens often live on plant surfaces in low numbers without causing immediate symptoms.
- They can travel long distances on moist air currents or be moved short distances on splashing rain and irrigation.
- When weather conditions are right, pathogen populations grow dramatically and cause disease.

## Management

- Most fungicide sprays will not protect the tree or shrub from a bacterial pathogen.
- Only sprays with copper or streptomycin as an active ingredient are effective in preventing bacterial diseases.

## Management of Leaf Spot Diseases of Trees and Shrubs

Leaf spot diseases will not seriously harm your plants, but there are things you can do that, when done together, can reduce the disease on the tree in following years.

Marssonia leaf spot

- Rake up and destroy fallen leaves before the first snowfall to eliminate locations where diseases can survive to re-infect the plant the following growing season.
- Do not overcrowd plants — use size at maturity as a spacing guide when planting.
- Prune trees or shrubs to increase light penetration and improve air circulation throughout the canopy.
- Wet conditions promote disease, so water trees at the base and be careful not to splash water on leaves. A drip or soaker hose works best for this. Avoid sprinklers.
- Reduce stress to your tree:
    - Water your tree throughout the growing season so that the top 6 to 8 inches of the soil is moist, especially during dry summer periods.
- Soil should be allowed to dry before watering again.

  - Maintain a 3- to 4-inch-deep layer of mulch around your tree.
- Do not mound the mulch around the trunk of the tree but lay a flat layer with at least a 2-inch space between the mulch and stem to allow for air movement.
- Annually reapply mulch and inspect to ensure levels are maintained.
- Do not fertilize trees and shrubs suffering from leaf spot diseases, unless it is recommended by a soil test to correct a nutrient deficiency.
- Fungicides are not necessary unless a tree has lost all of its leaves several years in a row.
- Fungicides are protective and need to be applied before symptoms appear on the leaves.
  - Proper timing of fungicide applications can vary depending on the biology of the disease.
  - High-pressure spraying equipment is needed in order to get complete coverage of the canopy of large trees.
  - Hire a professional arborist to treat leaf spot diseases in large trees.

# Soybean Rust

Soybean rust disease is caused by one of two different fungi, Phakopsora pachyrhizi and Phakopsora meibomiae. P. meibomiae, also called the New World type of soybean rust, is a weaker pathogen that is found in small areas of the Western Hemisphere.

P. pachyrhizi, called Asian or Australasian soybean rust, on the other hand, is far more virulent. First reported in Japan in 1902, the disease was only found in tropical to semitropical regions of Asia and Australia.

## Symptoms and Signs

The first symptoms of soybean rust caused by Phakopsora pachyrhizi begin as very small brown or brick-red spots on leaves. Symptoms caused by P. meibomiae are similar to those of P. pachyrhizi, but this topic will focus on P. pachyrhizi because most of the research and observations have been made with this species. In the field, these spots usually begin in the lower canopy at or after flowering, although seedlings can be infected under certain circumstances. Often the first lesions appear toward the base of the leaflet near the petiole and leaf veins. This part of the leaflet probably retains dew longer, making conditions more favorable for infection. Lesions remain small (2-5 mm in diameter), but increase in number as the disease progresses. Pustules, called uredinia, form in these lesions, mostly on the lower leaf surface, and they can produce many urediniospores. The raised pustules can be seen with the unaided eye, especially when sporulating. Even though the lesions are small, each lesion often has several pustules (uredinia). Lesions can be completely covered in urediniospores when the pustules are active. Soybean rust urediniospores are pale yellow-brown to colorless, with an echinulate (short spines) surface ornamentation. This coloration is different from many other rust pathogens whose spores are often reddish-brown (rust colored). Germination of P. pachyrhizi urediniospores occurs through an equatorial (central) pore, producing a germ tube that ends in an appressorium, which the fungus uses to penetrate the host directly or through a stoma.

As more and more lesions form on a leaflet, the affected area begins to yellow, and eventually the leaflet falls from the plant. While soybean rust usually begins in the lower canopy, it quickly progresses up the plant until all of the leaves have some level of disease. Severely diseased plants may become completely defoliated. The loss of effective leaf tissue results in yield reductions from both fewer and smaller seed. Yield losses as high as 30 to 80% have been reported, but the amount of loss depends on when the disease begins and how rapidly it progresses. Besides leaves, soybean rust can also appear on petioles, stems, and even cotyledons, but most rust lesions occur on leaves.

Lesions may be either tan or red-brown. Tan lesions have many pustules that produce numerous urediniospores. Red-brown lesions, thought to be a moderate resistance reaction, have only a few pustules that produce only a few urediniospores. This lesion type depends on the strain of the pathogen, and may appear on the same leaf with tan lesions, or tan lesions may turn red-brown with age. Symptoms and signs on other hosts, such as kudzu, are similar, although lesion size may differ.

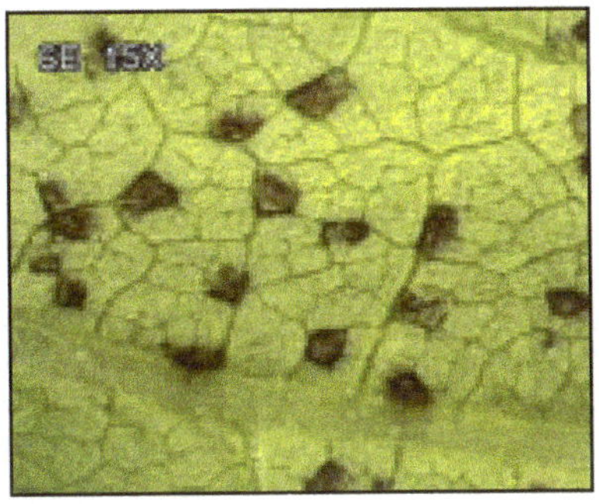
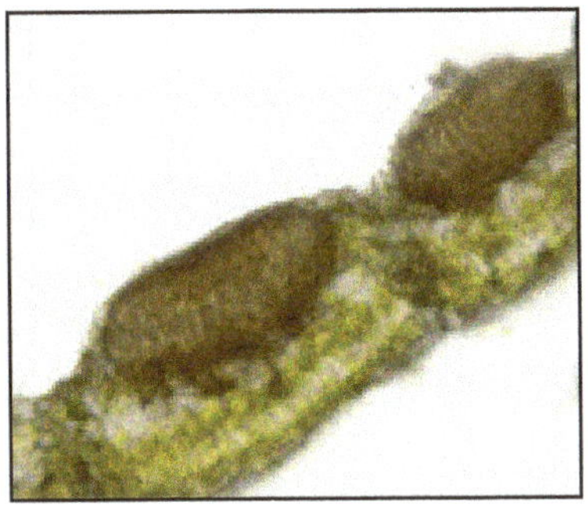

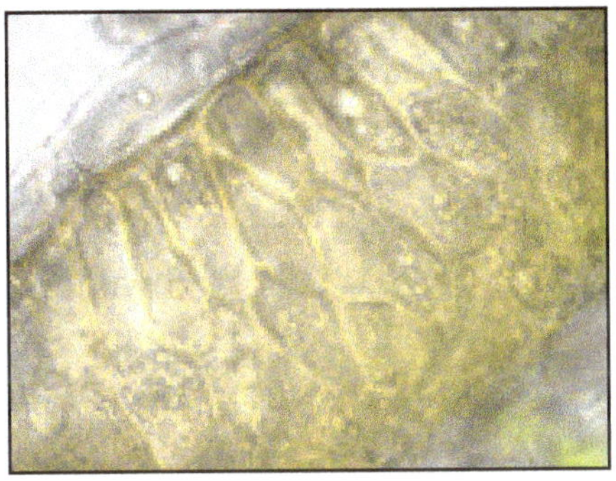

As pustules age, they may turn black. This is caused by the formation of a layer of teliospores in the pustules, turning pustules from uredinia into telia. Teliospores have two functions: survival of the fungus in the absence of a living host (overseasoning) and sexual reproduction. The thick walls of the teliospores protect the fungus from the environment and attack by other organisms. In rusts, the teliospores germinate forming a basidium and four basidiospores during which sexual recombination occurs. Germination of P. pachyrhizi teliospores has been observed only in the laboratory and does not seem to make a significant contribution to the perpetuation of this disease in the field.

## Pathogen Biology

Except for a few minor characteristics, the two fungi appear morphologically identical, but P. pachyrhizi is much more aggressive on soybean than P. meibomiae. To date, P. meibomiae has not been documented to cause significant yield losses in Central and South America. The two species can be distinguished by using DNA analysis protocols. Like other rusts, the soybean rust pathogens are obligate parasites that require a living host to grow and reproduce. They can survive away from the host as urediniospores for only a few days under natural conditions.

Both soybean rust pathogens, to the best of our knowledge, produce only two types of spores: urediniospores and teliospores. This contrasts with other rusts, which can have up to five spores stages (for example, wheat stem rust). For soybean rust, like most rusts, the uredinial stage is the repeating stage. This means that urediniospores can infect the same host on which they were produced (soybean) during the same season. Epidemics can develop quickly from only a few pustules because spore-producing pustules are produced in as little as 7 to 10 days after infection, and each pustule can produce hundreds of urediniospores.

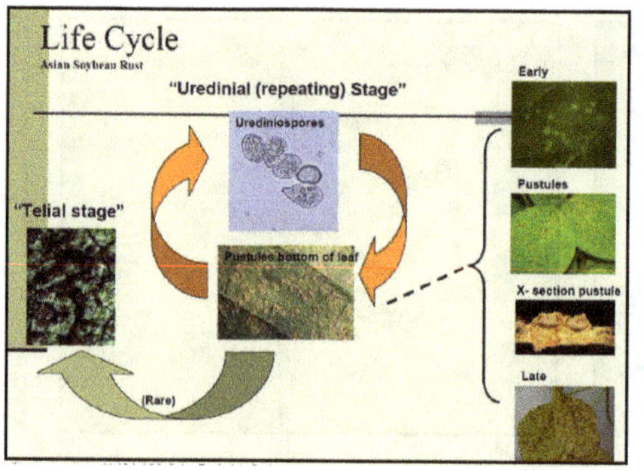


Teliospores are produced in old lesions, but they do not appear to germinate in nature, and no alternate host, nor aecia or spermogonia are known. Without germination of teliospores, sexual reproduction cannot take place. Lack of sexual reproduction should limit variability of the rust fungus, but nevertheless there is substantial variability in P. pachyrhizi with respect to virulence. This has limited the use of single genes for resistance in soybean, because in a short time new isolates of the pathogen arise that overcome the resistance gene. It is not known how this variability originates in P. pachyrhizi. Wheat stripe rust, Puccinia striiformis, has a similar life cycle as P. pachyrhizi with no functioning telial stage and therefore no sexual reproduction, but has many races. It may be that each resistance gene is so specific that a single mutation in the right gene of the fungus allows it to be virulent on hosts with the new resistance gene.

## Epidemiology

Soybean rust epidemics begin with the arrival of airborne inoculum (urediniospores). This pathogen is unique among rusts because it has many alternative hosts, which may serve as sources of inoculum. Alternative hosts are other plants that can become infected with the same pathogen, but are not required to complete the pathogen's life cycle. The principal host, that is needed for a pathogen to complete its life cycle. In frost-free areas, such as South America, Central America, the Caribbean basin, southern Texas, and Florida, the inoculum source could be nearby on volunteer soybean plants, kudzu, or some other alternative host. In areas that experience frost, such as the Midwestern United States, inoculum must be blown in from over-wintering sources that may be hundreds of miles away. Re-introduction of obligate pathogens into a distant region occurs with several other diseases, such as wheat stem rust and downy mildews, e.g. blue mold on tobacco. Because spores of P. pachyrhizi are sensitive to ultraviolet radiation, long distance movement of these rust spores probably occurs in storm systems where clouds protect the spores from the sun.

Table: Known hosts of Asian soybean rust caused by Phakopsora pachyrhizi. Information courtesy of Kent Smith, USDA/ARS:

| COMMON NAME | SCIENTFIC NAME |
|---|---|
| Bean, Common, Dry (field, kidney, navy, pinto) * | Phaseolus vulgaris var. vulgaris |
| Bean, Common, Succulent (garden, green, snap, and wax) * | Phaseolus vulgaris var. vulgaris |
| Bean, Fava or Broadbean | Vicia faba |
| Bean, Lablab or Hyacinth * | Lablab purpureus |

| Bean, Lima * | Phaseolus lunatus var. lunatus |
|---|---|
| Bean, Mung * | Vigna radiata |
| Bean, Scarlet Runner * | Phaseolus coccineus |
| Bean, Winged or Goa | Psophocarpus tetragonolobus |
| Bean, Yam * | Pachyrhizus ahipa, P. erosus |
| Blackeyed Pea, Cowpea or Yardlong Bean * | Vigna unguiculata |
| Calopo | Calopogonium mucunoides |
| Clover; Alyce or Oneleaf | Alysicarpus vaginalis |
| Clover, Crimson | Trifolium incarnatum |
| Clover, Hop | Trifolium aureum |
| Clover, Lappa | Trifolium lappaceum |
| Clover, White | Trifolium repens |
| Crotalaria * | Crotalaria anagyroides, C. spectabilis |
| Crownvetch | Securigera varia |
| Fenugreek | Trigonella foenum-graicum |
| Florida Beggarweed* | Desmodium tortuosum |
| Kudzu * | Pueraria montana var. lobata |
| Lespedeza | Lespedeza spp., Kummerowia striata, K. stipulaceae |
| Lupines | Lupinus spp. |
| Medic | Medicago spp. |
| Milk Vetch | Astragalus cicer, A. glycyphyllos |
| Pea, garden and field | Pisum sativum |
| Peatree or Colorado River Hemp (Sesbania) | Sesbania exaltata |
| Pigeon Pea * | Cajanus cajan |
| Siratro * | Macroptilium atropurpureum |
| Soybean (including edamame) * | Glycine max |
| Swordbean * | Canavalia gladiata |
| Trefoil | Lotus spp. |
| Urd or Black Gram * | Vigna mungo |
| Wild Soybean * | Neonotonia wightii |
| Woolypod Vetch | Vicia villosa subsp. varia |
| Yellow Sweet Clover | Melilotus officinalis |
| * Includes field observations of infection, in addition to infection resulting from artificial inoculation | |

Once viable spores have landed on the leaf surface of a suitable host, infection and subsequent epidemic development are dependent on environmental conditions. Generally, infection occurs when leaves are wet and temperatures are between 8° C and 28° C, with an optimum of 16° C to 28° C. At 25° C, some infection occurs in as little as 6 hours of leaf wetness, but 12 hours are optimal. After infection, lesions and pustules with urediniospores can appear within 7 or 8 days, and the next infection cycle is set to begin. This short life cycle means that, under the right conditions, soybean rust epidemics can quickly build up from almost undetectable levels to very high levels. Soybean rust epidemics can progress from below detectable levels to defoliation within a month. Epidemics may seem to progress even faster than that, because early infections occur in the lower canopy and are hard to find. Besides the environment, plant age affects soybean rust epidemics. Usually, rust lesions are not found on soybean until flowering, unless there are high inoculum levels early in the season. This may be due to greater susceptibility of plants to rust as the host enters the reproductive stages, it may be because in lower parts of the canopy spores are more protected from UV radiation, or it may be because conditions in the canopy become more humid as the canopy closes. In any event, lesions can form at any growth stage, but major increases in disease do not occur until after flowering.

## Disease Management

There are three basic management tactics that can play a role in reducing soybean rust epidemics: fungicides, genetic resistance, and cultural practices. At present, fungicides are the only highly effective tactic, but long-term management will probably depend more on resistance, in combination with fungicides and changes in cultural practices.

At present, the most effective means of managing soybean rust is the use of fungicides. However, to be effective, selecting the right fungicide and applying it at the right time are crucial. Several fungicides are registered in the US for soybean rust control, and most can be classified into three groups: chloronitriles, strobilurins, and triazoles. Chlorothalonil is the one chloronitrile fungicide registered for soybean rust control. Its protectant mode of action affects many biochemical pathways in the pathogen, but it is not taken up by the plant, not even by the cuticle. As a result, it is more subject to weathering than the strobilurins or the triazoles and complete coverage of the leaf surface is critical. To be effective, chlorothalonil may need to be reapplied several times if new growth or weathering occurs.

Strobilurin fungicides are modeled after a natural antifungal compound, strobilurin, produced by certain mushrooms. Strobilurins (also known as QoI fungicides) inhibit mitochondrial respiration in the pathogen. Strobilurins are typically absorbed by the cuticle, and act as protectant fungicides. A protectant fungicide prevents infections from taking place, but it has little effect on disease development once infection has occurred. Therefore, to be effective, protectants like the strobilurins must be applied before infection occurs. Depending on the rate applied, strobilurins are effective for up to 2 weeks after an application, but they will not protect newly developing leaves. Strobilurins control a broad range of soybean pathogens.

Triazoles inhibit sterol production, which disrupts cell membrane function in the pathogen. Triazoles are absorbed and translocated upward in the plant. While they generally do not prevent infection, the triazoles can kill the fungus in the plant and prevent pustules and spores from forming. The extent to which these chemicals are translocated depends on the triazole, but all of them move up the plant into new growth to one degree or another. Still, systemicity of triazoles in plants is incomplete and does not approach the level of systemicity associated with certain herbicides or insecticides. Triazoles are effective for 3 or 4 weeks after application and give some protection to new growth. While highly effective against rust, the triazoles are not as effective as the strobilurins against other soybean pathogens. Some fungicide products (premixes) contain both a triazole and a strobilurin. The premixes provide protection against a broader range of pathogens and reduce the possibility of pathogens developing resistance to either product.

The number of applications required for disease control depends on the compounds used, when the rust epidemic starts, and the favorability of the weather conditions. Even with triazoles, which are effective for the longest period of time, two applications are often needed to control soybean rust. In some locations in Brazil, high levels of inoculum early in the season result in rust epidemics starting well before flowering, thus forcing growers to make as many as five fungicide applications in order to control the disease. Such early disease onset and the early need for fungicide application are unlikely in most of the US. However, rust could start as early as flowering (R1) and require an additional spray before harvest. It is generally felt that, once the plants reach the R6 growth stage (when seeds have filled the pod), most of the yield has been achieved and controlling rust beyond that point is not economical. One concern with multiple applications of the same fungicide is the development of fungicide resistant pathogen strains. While fungicide resistance in P. pachyrhizi has not been reported, other fungal pathogens may be affected and growers should try not to spray the same fungicide consecutively. Fungicide labels may restrict the number of times a particular compound or class of compounds can be applied within a season to reduce the possibility of resistance developing.

The key to effective control of soybean rust with fungicides is application timing. This is especially important in areas of the US where the soybean rust pathogen must be reintroduced each year. The introduction or reintroduction will probably occur at different times in different years or not at all in some years. All of the fungicides, even the systemic triazoles, are most effective when applied just before the rust epidemic starts in the field. From tests in South America, if disease incidence reaches 10% in the lower canopy before the first application, fungicides will not completely control soybean rust, and some yield loss will result if weather conditions are favorable. Such low levels of disease are difficult to detect so growers need an early warning system that predicts the onset of disease early enough so that they have time to apply fungicides to all of their fields. Application decisions, equipment, and technique can all greatly impact the level of rust control achieved.

At present, the most reliable early detection method is the use of "sentinel plots." These are small plots (primarily soybean, but kudzu or another susceptible host also may be used) planted several weeks before the commercial crop, and often use early maturing cultivars. Both the early planting and the early maturity of the cultivars results in the sentinel plots flowering 1 to 3 weeks before the commercial crop. Since soybean rust usually develops after flowering, the disease can be observed in these sentinel plots a week or two before being found in adjacent commercial fields. This early warning gives growers in the area time to apply a protective fungicide treatment.

Another method of early detection of soybean rust is spore trapping, in which two strategies are being assessed. One traps windblown spores on glass slides coated with petroleum jelly. The spores are examined microscopically, and the presence and number of soybean rust-like spores noted. At this time, microscopic examination can only identify spores that resemble the soybean rust pathogen because it is not currently possible to identify P. pachyrhizi with certainty by simply examining the urediniospores. More conclusive identification of urediniospores of P. pachyrhizi is being developed by using labeled antibodies and polymerase chain reaction (PCR) protocols.

The other spore trapping approach involves collecting and filtering rainwater and then uses PCR to determine the presence of P. pachyrhizi on the filters. It is thought that long-distance spread of urediniospores occurs when storms pick up the spores and then deposit them in rainwater at distant locations. Because this technique uses species-specific molecular markers, positive findings are thought to be more reliable. In 2005 and 2006, both air and rain sampling found P. pachyrhizi or P. pachyrhizi-like spores over a wide area, far from where soybean rust was active. While neither approach can determine if the spores arrived alive, they do indicate that this pathogen has the potential to spread widely and quickly.

## Genetic Resistance

Soybean plants respond to infection by P. pachyrhizi by producing either tan, red-brown, or no lesions at all. Tan lesions produce many pustules with many spores. Red-brown lesions produce a few pustules with limited spore production, and no pustules or spores are produced where no lesions are formed. It is thought that these responses represent susceptible, moderate, or highly resistant reactions, respectively. High levels of resistance are usually associated with one or a few dominant genes. There are four known dominant genes for resistance to soybean rust, Rpp1 through Rpp4. While these dominant genes confer high levels of resistance and are relatively easy to incorporate into new soybean cultivars, they are not effective against all races of P. pachyrhizi. Deployment of varieties with new resistance genes is usually followed in a few years by the emergence of races of P. pachyrhizi that are virulent on them. This high degree of variability in the soybean rust pathogen is common in many rusts and requires the frequent discovery and incorporation of new sources of resistance. Currently, isolates of P. pachyrhizi exist that are virulent on each of the four known genes for resistance.

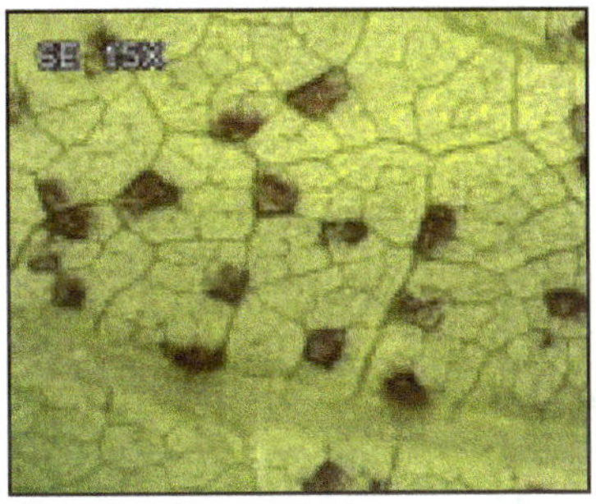

Another approach is the use of moderate resistance. Moderate resistance is usually conferred by a number of genes, each contributing a little to the overall resistance of the cultivar. This type of resistance often is effective against all races of a pathogen, but it is more difficult to incorporate into cultivars and does allow some disease and yield loss. Moderately resistant cultivars have been developed in Asia, but adapted varieties with this type of resistance are not yet available in the US or South America. Ultimately, moderate resistance may be used in combination with cultural practices and fungicides when needed.

## Cultural Practices

There are several cultural practices that may help manage soybean rust. In most areas of the US where rust must be introduced each year for an epidemic to occur, changing planting and harvest dates may avoid disease. Planting early with an early maturing cultivar may avoid the rust until the crop has either been harvested or is so far along that the disease will have little impact on yield. Planting dates may also be delayed so that the vulnerable reproductive period occurs during dry conditions that do not favor rust. In areas where the weather is marginal for rust development, wider row spacing along with lower plant populations may hasten canopy drying, thus reducing the dew period enough to prevent or at least slow disease development. It may also allow better fungicide penetration into the canopy, increasing the effectiveness of chemical control. Research is needed to confirm this. However, because the more open canopy provides less weed suppression, weed problems may be more severe with this strategy, and this method is unlikely to affect rust significantly if weather conditions are very favorable for the disease. Adjusting soil fertility, particularly potassium and phosphorus levels, may help increase disease resistance, but there is little research in this area yet. While it is unlikely that cultural control measures alone will be enough to control soybean rust, they may increase the effectiveness of host resistance or fungicide applications.

# Mosaic Virus Disease

Mosaic is a plant disease caused by various strains of several hundred viruses. A number of economically important crops are susceptible to mosaic infections, including tobacco, cassava, beet, cucumber, and alfalfa. Tulip mosaic virus "breaks" tulip and lily flowers, causing attractive and colorful streaking; this rare, unusual effect drove some of the frenzy of Tulip Mania in the 17th century.

Mosaic symptoms are variable but commonly include irregular leaf mottling (light and dark green or yellow patches or streaks). Leaves are commonly stunted, curled, or puckered; veins may be lighter than normal or banded with dark green or yellow. Plants are often dwarfed, with fruit and flowers fewer than usual, deformed, and stunted. Mosaic symptoms may be masked or latent, especially at temperatures above 27° C (81° F), and are sometimes confused with nutrient deficiency or herbicide injury. The causal viruses are spread by aphids and other insects, mites, fungi, nematodes, and contact; pollen and seeds can carry the infection as well.

Mosaic can be avoided by using virus-free seeds and plants, growing resistant varieties, separating new from old plantings, rotating annuals, and observing stringent sanitation and pest-control measures.

## Control and Prevention

### How to Control Mosaic Viruses

Once plants are infected, there are no controls. Remove all the infected plants and destroy them. Also, be sure to disinfect your gardening tools.

### Prevent Mosaic Viruses

- Plant resistant plants when available in your garden. Resistant varieties of tomatoes have yet to be developed for cucumber mosaic virus, but tomatoes that are resistant to tobacco mosaic virus may have some slight resistance to cucumber mosaic virus as well.
- Mosaic viruses are mostly spread by insects, especially aphids and leafhoppers. You can try covering your plants with a floating row cover or aluminum foil mulches to prevent these insects from infecting your plants. Look at our other tips for controlling aphids.
- Control your weeds. Some types may serve as hosts for the disease.
- To avoid tobacco mosaic virus, soak seeds in a 10 percent bleach solution before planting and avoid handling tobacco near plants.

# Potato Scab

Common scab of potatoes is a soil-borne disease which is caused by the bacteria-like organism Streptomyces scabies. This organism attacks the stems, stolons and roots of the potato, and more importantly, young, rapidly growing tubers, stimulating the growth of unsightly corky tissue. A severe infection can reduce the marketable yield and can damage the eyes of seed potatoes. It can also greatly reduce the market value of the crop.

Potato tubers infected with Streptomyces scabies.

## Symptoms

The scab organism varies greatly. This accounts for different symptoms and the range of conditions under which the potato can be damaged. Affected tubers may show either shallow or deep scabs or a combination of both.

A shallow or surface scab is defined as a superficial, roughened or russeted area on the tuber.

Slight protuberances with depressed centers may form and are covered with a small amount of corky tissue. A deep or pitted scab consists of lesions (areas of diseased tissue) which may be 1 mm to 10 mm deep, roughly circular, up to 10 mm in diameter and surrounded by corky tissue.

## Conditions Favoring Development of the Disease

Soil reaction (pH) and soil moisture seem to be the most important factors affecting seasonal incidence and regional distribution of scab, but there are other factors which are not clearly understood.

Common scab may be particularly severe when potatoes are grown in neutral or alkaline soils (pH 7.0 and above). Some races of the organism can damage tubers at a pH of 5.0 or below, but it is generally accepted that the disease is usually controlled satisfactorily if potato soils are kept at pH 5.0 to 5.2. Within the range of pH 5.2 to 8.0 the severity of the disease increases with increasing alkalinity.

Warm and dry soil conditions, particularly during the critical period of susceptibility to scab (two and five weeks after tubers start to form), can markedly increase the incidence of the disease.

## Disease Management

### Use of Resistant Varieties

The use of resistant varieties in soils where the problem recurs is the only economical means of long-term control. No popular variety is resistant but the early types (Red Pontiac, Kennebec and Sequoia) are more susceptible than the medium-to late-maturing types (Sebago and Coliban).

### Rotation

Long rotations of three to five years, preferably with legumes, but excluding beets, carrots, parsnips and fleshy-rooted crucifers, are useful in reducing the severity of the disease.

### Adequate Moisture and Timing of Irrigation

Dry soil at the time when tubers start to form, and for the five weeks after that, will increase the chances of common scab occurring. Irrigate regularly at this time and maintain moisture at field capacity.

### Soil Management

- Avoid excessive use of lime, fresh animal manures and wood ash. If lime is necessary, do not raise the soil pH above 5.2; ideally, use dolomitic lime after potatoes in the rotation.
- The lowering of soil pH to between 5.0 and 5.2 with applications of sulphur has proved useful in reducing the level of scab in some soils of high pH.

Use acid-producing fertilisers and use ammonium sulphate as a source of nitrogen.

### Combined Control Measures

It is important to realize that no one control measure is as effective as a combination of several. For example, controlled irrigation combined with sulphur treatment results in better control than either treatment used on its own.

## References


- Battillo, J. (2018). "The role of corn fungus in Basketmaker II diet: A paleonutrition perspective on early corn farming adaptations". Journal of Archaeological Science: Reports. 21: 64–70. doi:10.1016/j.jasrep.2018.07.003
- Anthracnose-disease, plant-problems: gardeningknowhow.com, Retrieved 09 July 2020
- Anthracnose, pest: almanac.com, Retrieved 11 April 2020
- Plant-disease, science: britannica.com, Retrieved 22, August 2020
- Banuett, F. (1995). "Genetics of Ustilago Maydis, A Fungal Pathogen that Induces Tumors in Maize". Annual Review of Genetics. 29 (1): 179–208. doi:10.1146/annurev.ge.29.120195.001143. PMID 8825473
- Shanmugasundaram, S.; Masuda, Ryoichi; Tsou, S.C.S.; Hong, T.L. (1991). Vegetable Soybean Research Needs for Production and Quality Improvement (PDF). Taipei: Asian Vegetable Research and Development Center. pp. 86–87. ISBN 9789290580478. Retrieved 6 February 2020

- What-is-damping-off-disease: gardenbetty.com, Retrieved 10 June 2020
- Molina, L; Kahmann, R (2007). "An Ustilago maydis gene involved in H2O2 detoxification is required for virulence". Plant Cell. 19 (7): 2293–2309. doi:10.1105/tpc.107.052332. PMC 1955693. PMID 17616735
- What-is-bacterial-canker, ornamental: gardeningknowhow.com, Retrieved 25 May 2020

# Chapter 4

# Plant Diseases: Resistance and Management

The reduction of pathogen growth in plants is known as disease resistance. Plant disease management aims to reduce the aesthetic and economic damage caused by plant disease. Plant disease resistance, disease resistance in fruits and vegetables, pest control and fungicide are the important topics covered in this chapter.

## Plant Disease Resistance

Plant disease resistance protects plants from pathogens in two ways: mechanisms and by infection-induced responses of the immune system. Relative to a susceptible plant, disease resistance is the reduction of pathogen growth on or in the plant, while the term disease tolerance describes plants that exhibit little disease damage despite substantial pathogen levels. Disease outcome is determined by the three-way interaction of the pathogen, the plant and the environmental conditions (an interaction known as the disease triangle).

Cankers caused Chestnut blight a disease that affects the chestnut tree

Defense-activating compounds can move cell-to-cell and systemically through the plant vascular system. However, plants do not have circulating immune cells, so most cell types exhibit a broad suite of antimicrobial defenses. Although obvious *qualitative* differences in disease resistance can be observed when multiple specimens are compared (allowing classification as "resistant" or "susceptible" after infection by the same pathogen strain at similar inoculum levels in similar environments), a gradation of *quantitative* differences in disease resistance is more typically observed between plant strains or genotypes. Plants consistently resist certain pathogens but succumb to others; resistance is usually pathogen species- or pathogen strain-specific.

## Background

Plant disease resistance is crucial to the reliable production of food, and it provides significant reductions in agricultural use of land, water, fuel and other inputs. Plants in both natural and cultivated populations carry inherent disease resistance, but this has not always protected them.

The late blight Irish potato famine of the 1840s was caused by the oomycete Phytophthora infestans. The world's first mass-cultivated banana cultivar Gros Michel was lost in the 1920s to Panama disease caused by the fungus Fusarium oxysporum. The current wheat stem, leaf, and yellow stripe rust epidemics spreading from East Africa into the Indian subcontinent are caused by rust fungi Puccinia graminis and P. striiformis. Other epidemics include Chestnut blight, as well as recurrent severe plant diseases such as Rice blast, Soybean cyst nematode, Citrus canker.

Plant pathogens can spread rapidly over great distances, vectored by water, wind, insects, and humans. Across large regions and many crop species, it is estimated that diseases typically reduce plant yields by 10% every year in more developed nations or agricultural systems, but yield loss to diseases often exceeds 20% in less developed settings, an estimated 15% of global crop production.

However, disease control is reasonably successful for most crops. Disease control is achieved by use of plants that have been bred for good resistance to many diseases, and by plant cultivation approaches such as crop rotation, pathogen-free seed, appropriate planting date and plant density, control of field moisture and pesticide use.

## Viral Disease Common Mechanisms

## Pre-formed Structures and Compounds

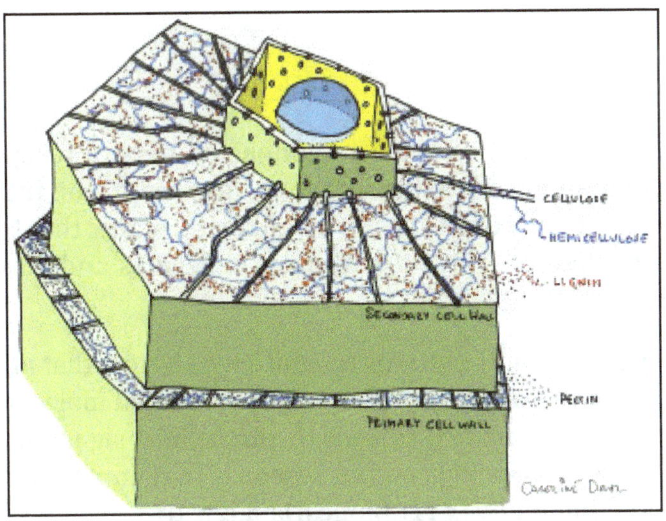


secondary plant wall

- Plant cuticle/surface.
- Plant cell walls.
- Antimicrobial chemicals (for example: glucosides, saponins).
- Antimicrobial proteins.
- Enzyme inhibitors.
- Detoxifying enzymes that break down pathogen-derived toxins.
- Receptors that perceive pathogen presence and activate inducible plant defences.

## Inducible Post-infection Plant Defenses

- Cell wall reinforcement (callose, lignin, suberin, cell wall proteins).
- Antimicrobial chemicals, including reactive oxygen species such as hydrogen peroxide or peroxynitrite, or more complex phytoalexins such as genistein or camalexin.
- Antimicrobial proteins such as defensins, thionins, or PR-1.
- Antimicrobial enzymes such as chitinases, beta-glucanases, or peroxidases.
- Hypersensitive response - a rapid host cell death response associated with defence mediated by "Resistance genes."(Bryant, Tracy 2008).

## Variable Resistance

Even in susceptible plants to obligate parasites the tissue resistance changes due to ontogeny and to influence of external conditions. This resistance can be measured by the value of redox potential of electron carriers, which is produced in the plant by enzymatic reactions associated with respiration. Electron carriers are water-soluble and are not oxidized by air oxygen. There is not free oxygen in the cells, all the oxidizations and reductions take place enzymatically. These reactions are highly specific for the plant species. The host and parasite have different electron carriers.

## Immune System

The plant immune system consists of two interconnected tiers of receptors, one outside and one inside the cell. Both systems sense the intruder, respond to the intrusion and optionally signal to the rest of the plant and sometimes to neighboring plants that the intruder is present. The two systems detect different types of pathogen molecules and classes of plant receptor proteins.

The first tier is primarily governed by pattern recognition receptors that are activated by recognition of evolutionarily conserved pathogen or microbial–associated molecular patterns (PAMPs or MAMPs, here P/MAMP). Activation of PRRs leads to intracellular signaling, transcriptional reprogramming, and biosynthesis of a complex output response that limits colonization. The system is known as PAMP-Triggered Immunity (PTI)"(JonesDangl2010).

The second tier (again, primarily), effector-triggered immunity (ETI), consists of another set of receptors, (nucleotide-binding)They operate within the cell, encoded by R genes. The presence of specific pathogen "effectors" activates specific NLR proteins that limit pathogen proliferation.

Receptor responses include ion channel gating, oxidative burst, cellular redox changes, or protein kinase cascades that directly activate cellular changes (such as cell wall reinforcement or antimicrobial production), or activate changes in gene expression that then elevate other defensive responses

Plant immune systems show some mechanistic similarities with the immune systems of insects and mammals, but also exhibit many plant-specific characteristics. Plants can sense the presence of pathogens and the effects of infection via activated by touch]. Rice Universi

## PAMP-triggered Immunity

PAMP-Triggered Immunity conserved molecules that inhabit multiple pathogen genera are classified as MAMPs by some researchers. The defenses induced by MAMP perception are sufficient to repel most pathogens. However, pathogen effector proteins are adapted to suppress basal defenses such as PTI.

## Effector Triggered Immunity

Effector Triggered Immunity (ETI) is activated by the presence of pathogen effectors. The ETI immune response is reliant on R genes, and is activated by specific pathogen strains. As with PTI, many specific examples of apparent ETI violate commoMost plant immune systems carry a repertoire of 100-600 different R genes that mediate resistance to various virus .Plant ETI often cause an apoptotic hypersensitive response.(Odds Rathjen2010)

## R Genes and R Proteins

Plants have evolved R genes (resistance genes) whose products allow recognition of specific pathogen effectors, either through direct binding or by recognition of the effector's alteration of a host protein. These virulence factors drove co-evolution of plant resistant genes to combat the pathogens' Avr (avirulent) genes. Many R genes encode NB-LRR proteins (nucleotide-binding/leucine-rich repeat domains, also known as NLR proteins or STAND proteins, among other names).

R gene products control a broad set of disease resistance responses whose induction is often sufficient to stop further pathogen growth/spread. Each plant genome contains a few hundred apparent R genes. Studied R genes usually confer specificity for particular pathogen strains. As first noted by Harold Flor in his mid-20th century formulation of the gene-for-gene relationship, the plant R gene and the pathogen Avr gene must have matched specificity for that R gene to confer resistance, suggesting a receptor/ligand interaction for Avr and R genes. Alternatively, an effector can modify its host cellular target (or a molecular decoy of that target) activating an NLR associated with the.

## Effector Biology

So-called "core" effectors are defined operationally by their wide distribution across the population of a particular pathogen and their substantial contribution to pathogen virulence. Genomics can be used to identify core effectors, which can then functionally define new R alleles, which can serve as breeding targets.

## Rna Silencing And Systemic Acquired Resistance Elicited By Prior Infections

Against viruses, plants often induce pathogen-specific gene silencing mechanisms mediated by RNA interference. T

Plant immune systems also can respond to an initial infection in one part of the plant by physiologically elevating the capacity for a successful defense response in other parts. Such responses include systemic acquired resistance, largely mediated by salicylic acid-dependent pathways, and induced systemic resistance, largely mediated.

Species-level resistance

In a small number of cases, plant genes are effective against an entire pathogen species, even though that species that is pathogenic on other genotypes of that host species. Examples include barley MLO against powdery mildew, wheat Lr34 against leaf rust and wheat Yr36 against stripe rust. An array of mechanisms for this type of resistance may exist depending on the particular gene and plant-pathogen combination. Other reasons for effective plant immunity can include a lack of coadaptation (the pathogen and/or plant lack multiple mechanisms needed for colonization and growth within that host species), or a particularly effective suite of pre-formed defenses.

## Signaling Mechanisms

### Perception of Pathogen Presence

Plant defense signaling is activated by pathogen-detecting receptors. The activated receptors frequently elicit reactive oxygen and nitric oxide production, calcium, potassium and proton ion fluxes, altered levels of salicylic acid and other hormones and activation of MAP kinases and other specific protein kinases. These events in turn typically lead to the modification of proteins that control gene transcription, and the activation of defense-associated gene expression.

In addition to PTI and ETI, plant defenses can be activated by the sensing of damage-associated compounds (DAMP), such as portions of the plant cell wall released during pathogenic infection. Many receptors for MAMPs, effectors and DAMPs have been discovered. Effectors are often detected by NLRs, while MAMPs and DAMPs are often detected by transmembrane receptor-kinases that carry LRR or LysM extracellular domains.

### Transcription Factors and the Hormone Response

Numerous genes and/or proteins as well as other molecules have been identified that mediate plant defense signal transduction. Cytoskeleton and vesicle trafficking dynamics help to orient plant defense responses toward the point of pathogen attack.

## Mechanisms of Transcription Factors and Hormones

Plant immune system activity is regulated in part by signaling hormones such as:

- Salicylic acid
- Jasmonic acid
- Ethylene

There can be substantial cross-talk among these pathways.

## Regulation by Degradation

As with many signal transduction pathways, plant gene expression during immune responses can be regulated by degradation. This often occurs when hormone binding to hormone receptors stimulates ubiquitin-associated degradation of repressor proteins that block expression of certain genes. The net result is hormone-activated gene expression. Examples:

- Auxin: binds to receptors that then recruit and degrade repressors of transcriptional activators that stimulate auxin-specific gene expression.
- Jasmonic acid: similar to auxin, except with jasmonate receptors impacting jasmonate-response signaling mediators such as JAZ proteins.
- Gibberellic acid: Gibberellin causes receptor conformational changes and binding and degradation of Della proteins.
- Ethylene: Inhibitory phosphorylation of the EIN2 ethylene response activator is blocked by ethylene binding. When this phosphorylation is reduced, EIN2 protein is cleaved and a portion of the protein moves to the nucleus to activate ethylene-response gene expression.

## Ubiquitin and E3 Signaling

Ubiquitination plays a central role in cell signaling that regulates processes including protein degradation and immunological response. Although one of the main functions of ubiquitin is to target proteins for destruction, it is also useful in signaling pathways, hormone release, apoptosis and translocation of materials throughout the cell. Ubiquitination is a component of several immune responses. Without ubiquitin's proper functioning, the invasion of pathogens and other harmful molecules would increase dramatically due to weakened immune defenses.

## E3 Signaling

The E3 Ubiquitin ligase enzyme is a main component that provides specificity in protein degradation pathways, including immune signaling pathways. The E3 enzyme components can be grouped by which domains they contain and include several types. These include the Ring and U-box single subunit, HECT, and CRLs. Plant signaling pathways including immune responses are controlled by several feedback pathways, which often include negative feedback; and they can be regulated by De-ubiquitination enzymes, degradation of transcription factors and the degradation of negative regulators of transcription.

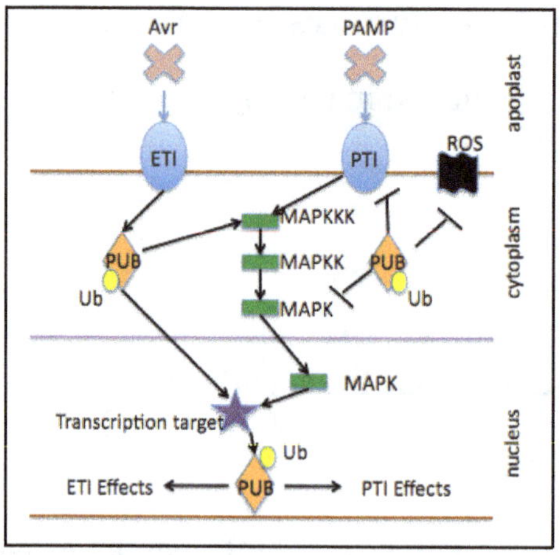


This image depicts the pathways taken during responses in plant immunity. It highlights the role and effect ubiquitin has in regulating the pathway.

## Plant Breeding for Disease Resistance

Plant breeders emphasize selection and development of disease-resistant plant lines. Plant diseases can also be partially controlled by use of pesticides and by cultivation practices such as crop rotation, tillage, planting density, disease-free seeds and cleaning of equipment, but plant varieties with inherent (genetically determined) disease resistance are generally preferred. Breeding for disease resistance began when plants were first domesticated. Breeding efforts continue because pathogen populations are under selection pressure for increased virulence, new pathogens appear, evolving cultivation practices and changing climate can reduce resistance and/or strengthen pathogens, and plant breeding for other traits can disrupt prior resistance. A plant line with acceptable resistance against one pathogen may lack resistance against others.

Breeding for resistance typically includes:

- Identification of plants that may be less desirable in other ways, but which carry a useful disease resistance trait, including wild strains that often express enhanced resistance.
- Crossing of a desirable but disease-susceptible variety to another variety that is a source of resistance.
- Growth of breeding candidates in a disease-conducive setting, possibly including pathogen inoculation. Attention must be paid to the specific pathogen isolates, to address variability within a single pathogen species.
- Selection of disease-resistant individuals that retain other desirable traits such as yield, quality and including other disease resistance traits.

Resistance is termed *durable* if it continues to be effective over multiple years of widespread use as pathogen populations evolve. "Vertical resistance" is specific to certain races or strains of a pathogen species, is often controlled by single R genes and can be less durable. Hoizontal or broad-spectrum resistance against an entire pathogen species is often only incompletely effective, but more durable, and is often controlled by many genes that segregate in breeding populations.

Crops such as potato, apple, banana and sugarcane are often propagated by vegetative reproduction to preserve highly desirable plant varieties, because for these species, outcrossing seriously disrupts the preferred traits. Asexual propagation. Vegetatively propagated crops may be among the best targets for resistance improvement by the biotechnology method of plant transformation to manage genes that affect disease resistance.

Scientific breeding for disease resistance originated with Sir Rowland Biffen, who identified a single recessive gene for resistance to wheat yellow rust. Nearly every crop was then bred to include disease resistance (R) genes, many by introgression from compatible wild relatives.

## GM or Transgenic Engineered Disease Resistance

The term GM ("genetically modified") is often used as a synonym of transgenic to refer to plants modified using recombinant DNA technologies. Plants with transgenic/GM disease resistance against insect pests have been extremely successful as commercial products, especially in maize and cotton, and are planted annually on over 20 million hectares in over 20 countries worldwide. Transgenic plant disease resistance against microbial pathogens was first demonstrated in 1986. Expression of viral coat protein gene sequences conferred virus resistance via small RNAs. This proved to be a widely applicable mechanism for inhibiting viral replication. Combining coat protein genes from three different viruses, scientists developed squash hybrids with field-validated, multiviral resistance. Similar levels of resistance to this variety of viruses had not been achieved by conventional breeding.

A similar strategy was deployed to combat papaya ringspot virus, which by 1994 threatened to destroy Hawaii's papaya industry. Field trials demonstrated excellent efficacy and high fruit quality. By 1998 the first transgenic virus-resistant papaya was approved for sale. Disease resistance has been durable for over 15 years. Transgenic papaya accounts for ~85% of Hawaiian production. The fruit is approved for sale in the U.S., Canada and Japan.

Potato lines expressing viral replicase sequences that confer resistance to potato leafroll virus were sold under the trade names NewLeaf Y and NewLeaf Plus, and were widely accepted in commercial production in 1999-2001, until McDonald's Corp. decided not to purchase GM potatoes and Monsanto decided to close their NatureMark potato business. NewLeaf Y and NewLeaf Plus potatoes carried two GM traits, as they also expressed Bt-mediated resistance to Colorado potato beetle.

No other crop with engineered disease resistance against microbial pathogens had reached the market by 2013, although more than a dozen were in some state of development and testing.

| Examples of transgenic disease resistance projects | | | | |
|---|---|---|---|---|
| **Publication year** | **Crop** | **Disease resistance** | **Mechanism** | **Development status** |
| 2012 | Tomato | Bacterial spot | R gene from pepper | 8 years of field trials |
| 2012 | Rice | Bacterial blight and bacterial streak | Engineered E gene | Laboratory |
| 2012 | Wheat | Powdery mildew | Overexpressed R gene from wheat | 2 years of field trials at time of publication |
| 2011 | Apple | Apple scab fungus | Thionin gene from barley | 4 years of field trials at time of publication |

| 2011 | Potato | Potato virus Y | Pathogen-derived resistance | 1 year of field trial at time of publication |
|---|---|---|---|---|
| 2010 | Apple | Fire blight | Antibacterial protein from moth | 12 years of field trials at time of publication |
| 2010 | Tomato | Multibacterial resistance | PRR from *Arabidopsis* | Laboratory scale |
| 2010 | Banana | Xanthomonas wilt | Novel gene from pepper | Now in field trial |
| 2009 | Potato | Late blight | R genes from wild relatives | 3 years of field trials |
| 2009 | Potato | Late blight | R gene from wild relative | 2 years of field trials at time of publication |
| 2008 | Potato | Late blight | R gene from wild relative | 2 years of field trials at time of publication |
| 2008 | Plum | Plum pox virus | Pathogen-derived resistance | Regulatory approvals, no commercial sales |
| 2005 | Rice | Bacterial streak | R gene from maize | Laboratory |
| 2002 | Barley | Stem rust | Resting lymphocyte kinase (RLK) gene from resistant barley cultivar | Laboratory |
| 1997 | Papaya | Ring spot virus | Pathogen-derived resistance | Approved and commercially sold since 1998, sold into Japan since 2012 |
| 1995 | Squash | Three mosaic viruses | Pathogen-derived resistance | Approved and commercially sold since 1994 |
| 1993 | Potato | Potato virus X | Mammalian interferon-induced enzyme | 3 years of field trials at time of publication |

## PRR Transfer

Research aimed at engineered resistance follows multiple strategies. One is to transfer useful PRRs into species that lack them. Identification of functional PRRs and their transfer to a recipient species that lacks an orthologous receptor could provide a general pathway to additional broadened PRR repertoires. For example, the *Arabidopsis* PRR *EF-Tu* receptor (EFR) recognizes the bacterial translation elongation factor *EF-Tu*. Research performed at Sainsbury Laboratory demonstrated that deployment of EFR into either *Nicotiana benthamiana*or *Solanum lycopersicum* (tomato), which cannot recognize *EF-Tu*, conferred resistance to a wide range of bacterial pathogens. EFR expression in tomato was especially effective against the widespread and devastating soil bacterium Ralstonia solanacearum. Conversely, the tomato PRR *Verticillium 1* (*Ve1*) gene can be transferred from tomato to *Arabidopsis*, where it confers resistance to race 1 Verticillium isolates.

## Stacking

The second strategy attempts to deploy multiple NLR genes simultaneously, a breeding strategy known as stacking. Cultivars generated by either DNA-assisted molecular breeding or gene transfer will likely display more durable resistance, because pathogens would have to mutate multiple effector genes. DNA sequencing allows researchers to functionally "mine" NLR genes from multiple species/strains.

The *avrBs2* effector gene from *Xanthomona perforans* is the causal agent of bacterial spot disease of pepper and tomato. The first "effector-rationalized" search for a potentially durable R gene followed the finding that *avrBs2* is found in most disease-causing *Xanthomonas* species and is required for pathogen fitness. The *Bs2* NLR gene from the wild pepper, *Capsicum chacoense*, was moved into tomato, where it inhibited pathogen growth. Field trials demonstrated robust resistance without bactericidal chemicals. However, rare strains of *Xanthomonas* overcame *Bs2*-mediated resistance in pepper by acquisition of *avrBs2* mutations that avoid recognition but retain virulence. Stacking R genes that each recognize a different core effector could delay or prevent adaptation.

More than 50 loci in wheat strains confer disease resistance against wheat stem, leaf and yellow stripe rust pathogens. The Stem rust 35 (*Sr35*) NLR gene, cloned from a diploid relative of cultivated wheat, *Triticum monococcum*, provides resistance to wheat rust isolate *Ug99*. Similarly, *Sr33*, from the wheat relative *Aegilops tauschii*, encodes a wheat ortholog to barley *Mla* powdery mildew–resistance genes. Both genes are unusual in wheat and its relatives. Combined with the *Sr2* gene that acts additively with at least Sr33, they could provide durable disease resistance to *Ug99* and its derivatives.

## Executor Genes

Another class of plant disease resistance genes opens a "trap door" that quickly kills invaded cells, stopping pathogen proliferation. Xanthomonas and Ralstonia transcription activator–like (TAL) effectors are DNA-binding proteins that activate host gene expression to enhance pathogen virulence. Both the rice and pepper lineages independently evolved TAL-effector binding sites that instead act as an executioner that induces hypersensitive host cell death when up-regulated. *Xa27* from rice and Bs3 and Bs4c from pepper, are such "executor" (or "executioner") genes that encode non-homologous plant proteins of unknown function. Executor genes are expressed only in the presence of a specific TAL effector.

Engineered executor genes were demonstrated by successfully redesigning the pepper *Bs3* promoter to contain two additional binding sites for TAL effectors from disparate pathogen strains. Subsequently, an engineered executor gene was deployed in rice by adding five TAL effector binding sites to the *Xa27* promoter. The synthetic *Xa27* construct conferred resistance against Xanthomonas bacterial blight and bacterial leaf streak species.

## Host Susceptibility Alleles

Most plant pathogens reprogram host gene expression patterns to directly benefit the pathogen. Reprogrammed genes required for pathogen survival and proliferation can be thought of as "disease-susceptibility genes." Recessive resistance genes are disease-susceptibility candidates. For example, a mutation disabled an *Arabidopsis* gene encoding pectate lyase (involved in cell wall degradation), conferring resistance to the powdery mildew pathogen *Golovinomyces cichoracearum*. Similarly, the Barley *MLO* gene and spontaneously mutated pea and tomato *MLO* orthologs also confer powdery mildew resistance.

*Lr34* is a gene that provides partial resistance to leaf and yellow rusts and powdery mildew in wheat. *Lr34* encodes an adenosine triphosphate (ATP)–binding cassette (ABC) transporter. The dominant allele that provides disease resistance was recently found in cultivated wheat (not in wild strains) and, like *MLO* provides broad-spectrum resistance in barley.

Natural alleles of host translation elongation initiation factors *eif4e* and *eif4g* are also recessive viral-resistance genes. Some have been deployed to control potyviruses in barley, rice, tomato, pepper, pea, lettuce and melon. The discovery prompted a successful mutant screen for chemically induced *eif4e* alleles in tomato.

Natural promoter variation can lead to the evolution of recessive disease-resistance alleles. For example, the recessive resistance gene *xa13* in rice is an allele of *Os-8N3*. *Os-8N3* is transcriptionally activated by*Xanthomonas oryzae pv. oryzae* strains that express the TAL effector *PthXo1*. The *xa13* gene has a mutated effector-binding element in its promoter that eliminates *PthXo1* binding and renders these lines resistant to strains that rely on *PthXo1*. This finding also demonstrated that *Os-8N3* is required for susceptibility.

Xa13/Os-8N3 is required for pollen development, showing that such mutant alleles can be problematic should the disease-susceptibility phenotype alter function in other processes. However, mutations in the *Os11N3* (OsSWEET14) TAL effector–binding element were made by fusing TAL effectors to nucleases (TALENs). Genome-edited rice plants with altered *Os11N3* binding sites remained resistant to *Xanthomonas oryzae pv. oryzae*, but still provided normal development function.

## Gene Silencing

RNA silencing-based resistance is a powerful tool for engineering resistant crops. The advantage of RNAi as a novel gene therapy against fungal, viral and bacterial infection in plants lies in the fact that it regulates gene expression via messenger RNA degradation, translation repression and chromatin remodelling through small non-coding RNAs. Mechanistically, the silencing processes are guided by processing products of the double-stranded RNA (dsRNA) trigger, which are known as small interfering RNAs and microRNAs.

## Host Range

Among the thousands of species of plant pathogenic microorganisms, only a small minority have the capacity to infect a broad range of plant species. Most pathogens instead exhibit a high degree of host-specificity. Non-host plant species are often said to express *non-host resistance*. The term *host resistance* is used when a pathogen species can be pathogenic on the host species but certain strains of that plant species resist certain strains of the pathogen species. The causes of host resistance and non-host resistance can overlap. Pathogen host range can change quite suddenly if, for example, the pathogen's capacity to synthesize a host-specific toxin or effector is gained by gene shuffling/mutation, or by horizontal gene transfer.

## Epidemics and Population Biology

Native populations are often characterized by substantial genotype diversity and dispersed populations (growth in a mixture with many other plant species). They also have undergone of plant-pathogen coevolution. Hence as long as novel pathogens are not introduced/do not evolve, such populations generally exhibit only a low incidence of severe disease epidemics.

Monocrop agricultural systems provide an ideal environment for pathogen evolution, because they offer a high density of target specimens with similar/identical genotypes.

The rise in mobility stemming from modern transportation systems provides pathogens with access to more potential targets.

Climate change can alter the viable geographic range of pathogen species and cause some diseases to become a problem in areas where the disease was previously less important.

These factors make modern agriculture more prone to disease epidemics. Common solutions include constant breeding for disease resistance, use of pesticides, use of border inspections and plant import restrictions, maintenance of significant genetic diversity within the crop gene pool and constant surveillance to accelerate initiation of appropriate responses. Some pathogen species have much greater capacity to overcome plant disease resistance than others, often because of their ability to evolve rapidly and to disperse broadly.

## Disease Resistance in Fruit and Vegetables

There are a number of lines of defence against pests (that, those animals that cause damage to the plants we grow) and diseases in the O, principal among these being the practice of good husbandry, creating healthy soil and ensuring high standards of garden hygiene. But no matter how diverse and healthy the garden eco-system may be, there will always be a degree of disease and pest presence. In many ways, some level of pathogen population in the garden can be not only acceptable but desirable as they are indicative of a generally healthful and diverse environment, and add to the overall robustness of the system as an immunity to such detrimental influences will build up, particularly in a balanced polycultural regime. Indeed, most of the plants we grow will tend to be selected because they are trouble free, and those that are more susceptible to attack will have fallen by the wayside over time. However, most farmers find it unacceptable that the food crops they grow are damaged by pests.

For these crops there has been considerable research and selective breeding carried out in order to find cultivars that are resistant or immune to pest and disease damage. Breeding for plant disease resistance generally has involved finding suitable genetic material amongst existing stocks or in the wild, which is then incorporated into commercial varieties.

### Example: the Apple

In the case of apples, in which research is being carried out in order to develop resistance to diseases such as apple scab (*Venturia inaequalis*), powdery mildew (*Podosphaera leucotricha*), orchard fireblight (*Erwinia amylovora*), woolly apple aphid (*Eriosoma lanigerum*) and collar rot (*Phytophthora cactorum*), the main sources of resistant material used in breeding programmes such as those being run by East Malling Research Station in England or Hortresearch in New Zealand are major gene resistances derived from crab-apples. The Vf gene for black spot resistance is derived from the ornamental crab-apple species *Malus floribunda*. Most black spot resistant cultivars developed around the world carry this gene, but there are some selections that carry the Vr (from M. pumila) or Vm (from M. micromalus) gene. Major gene resistances to powdery mildew are derived from M. robusta (Pl1) and M. zumi (Pl2), and the apple cultivar *Northern Spy* has a long-standing reputation for its major gene resistance to woolly apple aphid. Since early this century this resistance has been used to develop woolly aphid resistant rootstocks such as MM.106 and M.793. Much later it was shown that the cultivar was also very resistant to collar rot and a useful breeding parent for this resistance.

### Resistance and Immunity

Some plants can tolerate the presence of large numbers of insects without being severely affected. This is not very satisfactory however as insects will still cause damage, and in fact further breeding and population expansion of the pest species is supported. Other varieties are less attractive to pests, but this can be difficult to sustain or demonstrate. The most valuable form of resistance is where the pest cannot survive as well on one variety as on another. In some cases this can actually make the plants immune to attack, as is the case with the lettuces *Avoncrisp* and *Avondefiance* which were bred at the Institute of Horticultural Research, Wellesbourne during the 1960s, which are fully resistant to lettuce root aphid (*Pemphigus bursarius*).

### Trade-off of Breeding for Resistance

Sometimes however there can be a *trade-off*, for those varieties which have increased immunity or resistance may be lacking in other qualities such as flavour, yield or quality. Celery resistant to the Fusarium fungus (*Fusarium oxysporum spp.*) may not succumb to this disease, but may also be unacceptably short, ribby and low yielding. Further, a cultivar that is resistant to one disease may be more susceptible to another that is equally important. A lettuce cultivar that is resistant to mosaic virus may be sensitive to corky root disease, whilst another that resists corky root may be vulnerable to downy mildew (*Brim lactic*). Another drawback to resistance is that depending on the host pathogen system, resistance is sometimes not long lasting as new pathogen strains quickly develop, and further research and breeding is constantly needed.

### Availability of Resistant Varieties

Resistant varieties are not available for all crops. For several of the most damaging plant diseases, such as Potato blight (*Phytophthora infestans*) and white rot (*Sclerotic cepivorum*) of the Allium family, no acceptable resistant cultivars are yet available. In addition, commercial seed companies and plant breeders rarely invest resources into developing resistant cultivars for more minor or speciality crops, which often tend to be those of greater interest to the organic grower.

In general it is probably fair to say that resistance will not fully guarantee total crop protection, but choosing resistant varieties should rather be considered as a part of an overall Integrated pest management strategy, especially against virus diseases. In particular they can be especially useful where the threat from specific pests and diseases is high.

## Pest Control

Pest control refers to the regulation or management of a species defined as a pest,and can be perceived to be detrimental to a person's health, the ecology or the economy. A practitioner of pest control is called an exterminator.

A crop duster applies low-insecticide bait that is targeted against western corn rootworms.

## History

Pest control is at least as old as agriculture, as there has always been a need to keep crops free from pests. In order to maximize food production, it is advantageous to protect crops from competing species of plants, as well as from herbivores competing with humans.

The conventional approach was probably the first to be employed, since it is comparatively easy to destroy weeds by burning them or plowing them under, and to kill larger competing herbivores, such as crows and other birds eating seeds. Techniques such as crop rotation, companion planting (also known as intercropping or mixed cropping), and the selective breeding of pest-resistant cultivars have a long history.

In the UK, following concern about animal welfare, humane pest control and deterrence is gaining ground through the use of animal psychology rather than destruction. For instance, with the urban red fox which territorial behaviour is used against the animal, usually in conjunction with non-injurious chemical repellents. In rural areas of Britain, the use of firearms for pest control is quite common. Airguns are particularly popular for control of small pests such as rats, rabbits and grey squirrels, because of their lower power they can be used in more restrictive spaces such as gardens, where using a firearm would be unsafe.

Chemical pesticides date back 4,500 years, when the Sumerians used sulfur compounds as insecticides. The Rig Veda, which is about 4,000 years old, also mentions the use of poisonous plants for pest control. It was only with the industrialization and mechanization of agriculture in the 18th and 19th century, and the introduction of the insecticides pyrethrum and derris that chemical pest control became widespread. In the 20th century, the discovery of several synthetic insecticides, such as DDT, and herbicides boosted this development. Chemical pest control is still the predominant type of pest control today, although its long-term effects led to a renewed interest in traditional and biological pest control towards the end of the 20th century.

## Causes

Sign in Ilfracombe, England designed to help control seagull presence

Many pests have only become a problem as a result of the direct actions by humans. Modifying these actions can often substantially reduce the pest problem. In the United States, raccoons caused a nuisance by tearing open refuse sacks. Many householders introduced bins with locking lids, which deterred the raccoons from visiting. House flies tend to accumulate wherever there is human activity and is virtually a global phenomenon, especially where food or food waste is exposed. Similarly, seagulls have become pests at many seaside resorts. Tourists would often feed the birds with scraps of fish and chips, and before long, the birds would rely on this food source and act aggressively towards humans.

Living organisms evolve and increase their resistance to biological, chemical, physical or any other form of control. Unless the target population is completely exterminated or is rendered incapable of reproduction, the surviving population will inevitably acquire a tolerance of whatever pressures are brought to bear - this results in an evolutionary arms race.

## Types of Pest Control

### Use of Pest-destroying Animals

Perhaps as far ago as 3000BC in Egypt, cats were being used to control pests of grain stores such as rodents. In 1939/40 a survey discovered that cats could keep a farm's population of rats down to a low level, but could not eliminate them completely. However, if the rats were cleared by trapping or poisoning, farm cats could stop them returning - at least from an area of 50 yards around a barn.

Ferrets were domesticated at least by 500 AD in Europe, being used as mousers. Mongooses have been introduced into homes to control rodents and snakes, probably at first by the ancient Egyptians.

## Biological Pest Control

Biological pest control is the control of one through the control and management of natural predators and parasites. For example: mosquitoes are often controlled by putting *Bt Bacillus thuringiensis* ssp. *israelensis*, a bacterium that infects and kills mosquito larvae, in local water sources. The treatment has no known negative consequences on the remaining ecology and is safe for humans to drink. The point of biological pest control, or any natural pest control, is to eliminate a pest with minimal harm to the ecological balance of the environment in its present form.

## Mechanical Pest Control

Mechanical pest control is the use of hands-on techniques as well as simple equipment and devices, that provides a protective barrier between plants and insects. For example: weeds can be controlled by being physically removed from the ground. This is referred to as tillage and is one of the oldest methods of weed control.

## Physical Pest Control

Dog control van, Rekong Peo, Himachal Pradesh, India

Physical pest control is a method of getting rid of insects and small rodents by removing, attacking, setting up barriers that will prevent further destruction of one's plants, or forcing insect infestations to become visual.

## Elimination of Breeding Grounds

Proper waste management and drainage of still water, eliminates the breeding ground of many pests.

Garbage provides food and shelter for many unwanted organisms, as well as an area where still water might collect and be used as a breeding ground by mosquitoes. Communities that have proper garbage collection and disposal, have far less of a problem with rats, cockroaches, mosquitoes, flies and other pests than those that don't.

Open air sewers are ample breeding ground for various pests as well. By building and maintaining a proper sewer system, this problem is eliminated.

Certain spectrums of LED light can "disrupt insects' breeding".

## Poisoned Bait

Poisoned bait is a common method for controlling rat populations, however is not as effective when there are other food sources around, such as garbage. Poisoned meats have been used for centuries for killing off wolves, birds that were seen to threaten crops, and against other creatures. This can be a problem, since a carcass which has been poisoned will kill not only the targeted animal, but also every other animal which feeds on the carcass. Humans have also been killed by coming in contact with poisoned meat, or by eating an animal which had fed on a poisoned carcass. This tool is also used to manage several caterpillars e.g. Spodoptera litura, fruit flies, snails and slugs, crabs etc.

## Field Burning

Traditionally, after a sugar cane harvest, the fields are all burned, to kill off any rodents, insects or eggs that might be in the fields.

## Hunting

Historically, in some European countries, when stray dogs and cats became too numerous, local populations gathered together to round up all animals that did not appear to have an owner and kill them. In some nations, teams of rat-catchers work at chasing rats from the field, and killing them with dogs and simple hand tools. Some communities have in the past employed a bounty system, where a town clerk will pay a set fee for every rat head brought in as proof of a rat killing.

## Traps

A variety of mouse traps and rat traps are available for mice and rats, including snap traps, glue traps and live catch traps.

## Pesticides

Spraying pesticides by planes, trucks or by hand is a common method of pest control. Crop dusters commonly fly over farmland and spray pesticides to kill off pests that would threaten the crops. However, some pesticides may cause cancer and other health problems, as well as harming wildlife.

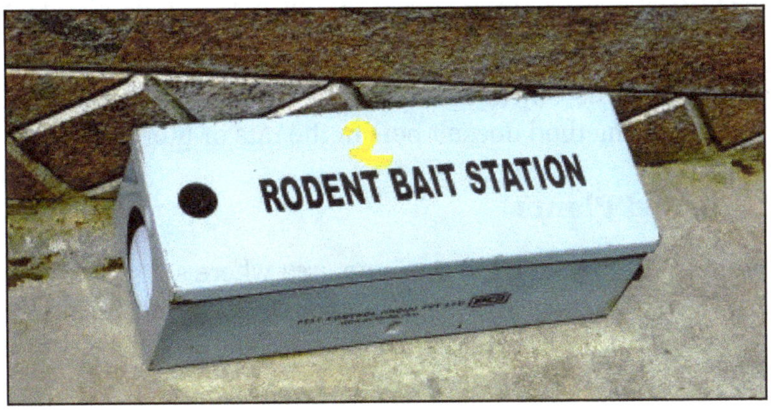


Rodent bait station, Chennai, India

## Space Fumigation

A project that involves a structure be covered or sealed airtight followed by the introduction of a penetrating, deadly gas at a killing concentration a long period of time (24-72hrs.). Although expensive, space fumigation targets all life stages of pests.

## Space Treatment

Residential & commercial building pest control service vehicle, Ypsilanti Township, Michigan

A long term project involving fogging or misting type applicators. Liquid insecticide is dispersed in the atmosphere within a structure. Treatments do not require the evacuation or airtight sealing of a building, allowing most work within the building to continue but at the cost of the penetrating effects. Contact insecticides are generally used, minimizing the long lasting residual effects. On August 10, 1973, the Federal Register printed the definition of Space treatment as defined by the U.S. Environmental Protection Agency (EPA):

> "The dispersal of insecticides into the air by foggers, misters, aerosol devices or vapor dispensers for control of flying insects and exposed crawling insects".

### Sterilization

Laboratory studies conducted with U-5897 (3-chloro-1,2-propanediol) were attempted in the early 1970s although these proved unsuccessful. Research into sterilization bait is ongoing.

In 2013, New York City tested sterilization traps in a $1.1 million study. The result was a 43% reduction in rat populations. The Chicago Transit Authority plans to test sterilization control in spring 2015. The sterilization method doesn't poison the rats or humans.

### Destruction of Infected Plants

Forest services sometimes destroy all the trees in an area where some are infected with insects, if seen as necessary to prevent the insect species from spreading. Farms infested with certain insects, have been burned entirely, to prevent the pest from spreading elsewhere.

### Natural Rodent Control

Example of House mouse infestation

Several wildlife rehabilitation organizations encourage natural form of rodent control through exclusion and predator support and preventing secondary poisoning altogether.

The United States Environmental Protection Agency agrees, noting in its Proposed Risk Mitigation Decision for Nine Rodenticides that "without habitat modification to make areas less attractive to commensal rodents, even eradication will not prevent new populations from recolonizing the habitat."

### Repellents

- Balsam fir oil from the tree *Abies balsamea* is an EPA approved non-toxic rodent repellent.
- *Acacia polyacantha* subsp. *campylacantha* root emits chemical compounds that repel animals including crocodiles, snakes and rats.

## Fungicide

Fungicides are biocidal chemical compounds or biological organisms used to kill fungi or fungal spores. A fungistatic inhibits their growth. Fungi can cause serious damage in agriculture, resulting in critical losses of yield, quality, and profit. Fungicides are used both in agriculture and to fight fungal infections in animals. Chemicals used to control oomycetes, which are not fungi, are also referred to as fungicides, as oomycetes use the same mechanisms as fungi to infect plants.

Fungicides can either be contact, translaminar or systemic. Contact fungicides are not taken up into the plant tissue and protect only the plant where the spray is deposited. Translaminar fungicides redistribute the fungicide from the upper, sprayed leaf surface to the lower, unsprayed surface. Systemic fungicides are taken up and redistributed through the xylem vessels. Few fungicides move to all parts of a plant. Some are locally systemic, and some move upwardly.

Most fungicides that can be bought retail are sold in a liquid form. A very common active ingredient is sulfur, present at 0.08% in weaker concentrates, and as high as 0.5% for more potent fungicides. Fungicides in powdered form are usually around 90% sulfur and are very toxic. Other active ingredients in fungicides include neem oil, rosemary oil, jojoba oil, the bacterium *Bacillus subtilis*, and the beneficial fungus *Ulocladium oudemansii*.

Fungicide residues have been found on food for human consumption, mostly from post-harvest treatments. Some fungicides are dangerous to human health, such as vinclozolin, which has now been removed from use. Ziram is also a fungicide that is thought to be toxic to humans if exposed to chronically. A number of fungicides are also used in human health care.

## Natural Fungicides

Plants and other organisms have chemical defenses that give them an advantage against microorganisms such as fungi. Some of these compounds can be used as fungicides:

- Tea tree oil
- Cinnamaldehyde
- Citronella oil
- Jojoba oil
- Nimbin
- Oregano oil
- Rosemary oil
- Monocerin
- Milk

Whole live or dead organisms that are efficient at killing or inhibiting fungi can sometimes be used as fungicides:

- *Bacillus subtilis*
- *Ulocladium oudemansii*

- Kelp (powdered dried kelp is fed to cattle to help prevent fungal infection)
- *Ampelomyces quisqualis*

## Resistance

Pathogens respond to the use of fungicides by evolving resistance. In the field several mechanisms of resistance have been identified. The evolution of fungicide resistance can be gradual or sudden. In qualitative or discrete resistance, a mutation (normally to a single gene) produces a race of a fungus with a high degree of resistance. Such resistant varieties also tend to show stability, persisting after the fungicide has been removed from the market. For example, sugar beet leaf blotch remains resistant to azoles years after they were no longer used for control of the disease. This is because such mutations often have a high selection pressure when the fungicide is used, but there is low selection pressure to remove them in the absence of the fungicide.

In instances where resistance occurs more gradually, a shift in sensitivity in the pathogen to the fungicide can be seen. Such resistance is polygenic – an accumulation of many mutations in different genes, each having a small additive effect. This type of resistance is known as quantitative or continuous resistance. In this kind of resistance, the pathogen population will revert to a sensitive state if the fungicide is no longer applied.

Little is known about how variations in fungicide treatment affect the selection pressure to evolve resistance to that fungicide. Evidence shows that the doses that provide the most control of the disease also provide the largest selection pressure to acquire resistance, and that lower doses decrease the selection pressure.

In some cases when a pathogen evolves resistance to one fungicide, it automatically obtains resistance to others – a phenomenon known as cross resistance. These additional fungicides are normally of the same chemical family or have the same mode of action, or can be detoxified by the same mechanism. Sometimes negative cross resistance occurs, where resistance to one chemical class of fungicides leads to an increase in sensitivity to a different chemical class of fungicides. This has been seen with carbendazim and diethofencarb.

There are also recorded incidences of the evolution of multiple drug resistance by pathogens – resistance to two chemically different fungicides by separate mutation events. For example, *Botrytis cinerea* is resistant to both azoles and dicarboximide fungicides.

There are several routes by which pathogens can evolve fungicide resistance. The most common mechanism appears to be alteration of the target site, in particular as a defence against single site of action fungicides. For example, Black Sigatoka, an economically important pathogen of banana, is resistant to the QoI fungicides, due to a single nucleotide change resulting in the replacement of one amino acid (glycine) by another (alanine) in the target protein of the QoI fungicides, cytochrome b. It is presumed that this disrupts the binding of the fungicide to the protein, rendering the fungicide ineffective. Upregulation of target genes can also render the fungicide ineffective. This is seen in DMI-resistant strains of *Venturia inaequalis*.

Resistance to fungicides can also be developed by efficient efflux of the fungicide out of the cell. *Septoria tritici* has developed multiple drug resistance using this mechanism. The pathogen had

5 ABC-type transporters with overlapping substrate specificities that together work to pump toxic chemicals out of the cell.

In addition to the mechanisms outlined above, fungi may also develop metabolic pathways that circumvent the target protein, or acquire enzymes that enable metabolism of the fungicide to a harmless substance.

## Fungicide Resistance Management

The fungicide resistance action committee (FRAC) has several recommended practices to try to avoid the development of fungicide resistance, especially in at-risk fungicides including *Strobilurins* such as azoxystrobin.

Products should not be used in isolation, but rather as mixture, or alternate sprays, with another fungicide with a different mechanism of action. The likelihood of the pathogen's developing resistance is greatly decreased by the fact that any resistant isolates to one fungicide will be killed by the other; in other words, two mutations would be required rather than just one. The effectiveness of this technique can be demonstrated by Metalaxyl, a phenylamide fungicide. When used as the sole product in Ireland to control potato blight (*Phytophthora infestans*), resistance developed within one growing season. However, in countries like the UK where it was marketed only as a mixture, resistance problems developed more slowly.

Fungicides should be applied only when absolutely necessary, especially if they are in an at-risk group. Lowering the amount of fungicide in the environment lowers the selection pressure for resistance to develop.

Manufacturers' doses should always be followed. These doses are normally designed to give the right balance between controlling the disease and limiting the risk of resistance development. Higher doses increase the selection pressure for single-site mutations that confer resistance, as all strains but those that carry the mutation will be eliminated, and thus the resistant strain will propagate. Lower doses greatly increase the risk of polygenic resistance, as strains that are slightly less sensitive to the fungicide may survive.

It is also recommended that where possible fungicides are used only in a protective manner, rather than to try to cure already-infected crops. Far fewer fungicides have curative/eradicative ability than protectant. Thus, fungicide preparations advertised as having curative action may have only one active chemical; a single fungicide acting in isolation increases the risk of fungicide resistance.

It is better to use an integrative pest management approach to disease control rather than relying on fungicides alone. This involves the use of resistant varieties and hygienic practices, such as the removal of potato discard piles and stubble on which the pathogen can overwinter, greatly reducing the titre of the pathogen and thus the risk of fungicide resistance development.

## References


- "Help WildCare Pursue Stricter Rodenticide Controls in California". wildcarebayarea.org/. Wild Care. Retrieved 28 February 2020
- Fred Baur. Insect Management for Food Storage and Processing. American Association of Cereal Chemists. ISBN 0-913250-38-4.

- "Pesticides". National Institute of Health Sciences. National Institute of Environmental Health. Retrieved 5 April 2020
- WOODY, TODD (September 20, 2010). "A Crop Sprouts Without Soil or Sunshine". nytimes.com. The New York Times. Retrieved 28 February 2020
- "Safer Rodenticide Products". epa.gov. USA Environment Protection Agency. March 2013. Retrieved 23 February 2020

# Chapter 5

## Plant Defense Mechanisms

The range of adaptations evolved by plants to improve their survival and reproduction are termed as plant defence mechanisms. This chapter thoroughly discusses plant tolerances and defense against herbivory, inducible plant defenses against herbivory and the use of Endophytic fungi by plants in their defense.

## Thigmonasty

*Mimosa pudica* in normal and touched state.

Thigmonasty or seismonasty is the nastic response of a plant or fungus to touch or vibration. Conspicuous examples of thigmonasty include many species in the leguminous subfamily Mimosoideae, active carnivorous plants such as Dionaea and a wide range of pollination mechanisms.

### Distinctive Aspects of Thigmonasty

Thigmonasty differs from thigmotropism in that nastic motion is independent of the direction of the stimulus. For example, tendrils from a climbing plant are thigmotropic because they twine around any support they touch, responding in whichever direction the stimulus came from. However, the shutting of a venus fly trap is thigmonastic; no matter what the direction of the stimulus, the trap simply shuts (and later possibly opens).

The time scales of thigmonastic responses tend to be shorter than those of thigmotropic movements because many examples of thigmonasty depend on pre-accumulated turgor or on bistable mechanisms rather than growth or cell division. Certain dramatic examples of rapid plant movement such as the sudden drooping of *Mimosa pudica* or the trapping action of *Dionaea* or *Utricularia* are fast enough to observe without time lapse photography; some take less than a second.

Speed is no clear distinction however; for example the re-erection of *Mimosa* leaves is nastic, but typically takes some 15 to 30 minutes, rather than a second or so. Similarly, re-opening of the *Dionaea* trap, though also nastic, typically takes days to complete.

Botanical physiologists have discovered signalling molecules called turgorins, that help mediate the loss of turgor. In species with the fastest response time, vacuoles are believed to provide temporary, high speed storage for calcium ions.

## Examples of Plants Exhibiting Thigmonasty

### Thigmonasty in the Asteraceae

Thigmonasty other than leaf closure occurs in various species of thistles. When an insect lands on a flower, the anthers shrink and rebound, loading the insect with pollen. The effect results from turgor changes in specialized, highly elastic cell walls of the anthers. Similar pollination strategy occurs in *Rudbeckia hirta.*

### Thigmonasty in the Droseraceae

The Venus Flytrap (*Dionaea muscipula*) presents a spectacular example of thigmonasty; when an insect lands on a trap formed by two curved lobes of a single leaf, the trap rapidly switches from an open to a closed configuration. Investigators have observed an action potential and changes in leaf turgor that accompany the reflex; they trigger the rapid elongation of individual cells. The common term for the elongation is acid growth although the process does not involve cell division.

### Thigmonasty in the Fabaceae

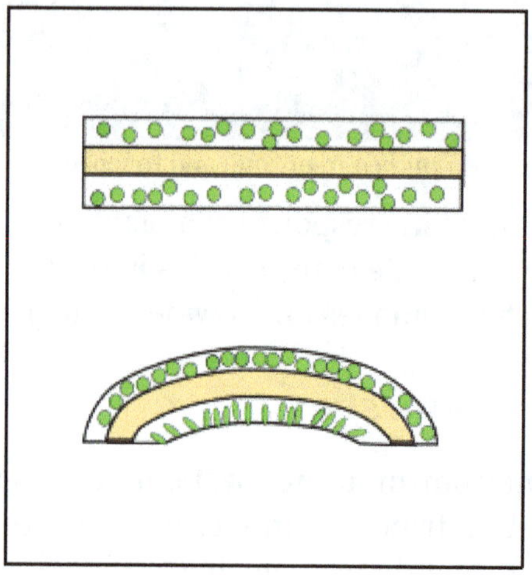

Pulvinus in extended and contracted position

*Mimosa pudica* is a plant with compound leaves that droop abruptly when stimulated. This is a classic example of thigmonastic action and has attracted detailed investigation. Contact or injury that causes leaflets to deform, will trigger an action potential. The action potential travels through the plant, initiating drooping of the leaflets as it passes. However, it cannot pass the pulvinus at the base of a petiole, and this prevents a local disturbance from causing all the leaves on the plant to collapse in response to trivial disturbances.

The pulvinus is a motor structure consisting of a rod of sclerenchyma surrounded by collenchyma. Such pulvini occur widely in the Fabaceae. In its extended position, the cells of the entire collar of collenchyma are distended with water. On receiving the action potential signal, the cells in the lower half of the pulvinus respond by expelling potassium and chlorine ions and taking up of calcium ions. This results in an osmotic gradient that draws water out of the affected cells, so that they temporarily shrink. This pulls the entire structure downward like a folding fan.

Many other Fabaceae react to touch with the same rapid leaf closure motion. These include the telegraph plant and the silk tree. The pea vine thigmonastically closes its leaves around a support. Catclaw Brier, a prairie mimosa, native to North America, shuts its leaves on contact. The plant is attractive to herbivores, and this behavior presumably provides protection against grazing.

### Thigmonasty in The Oxalidaceae

Sensitive leaves also occur in plants of the wood sorrel family. Examples include many species of *Oxalis*, *Biophytum sensitivum*, and Averrhoa carambola (the plant which produces starfruit).

### Other Forms

Some fungi exhibit trap closure similar to the venus fly trap. Mycologists have discovered action potentials in fungi but it is not currently clear whether they have any significance to thigmonastic behavior.

## Raphide

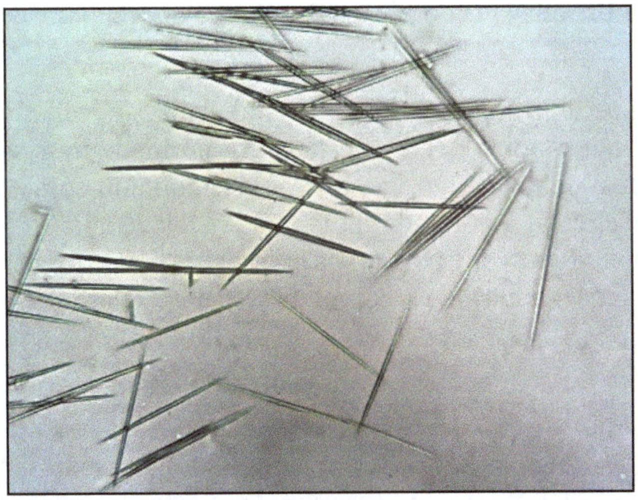

Raphides in *Epipremnum* Devil's ivy (600x magnification)

Raphides are needle-shaped crystals of calcium oxalate as the monohydrate or calcium carbonate as aragonite, found in more than 200 families of plants. Both ends are needle-like, but raphides tend to be blunt at one end and sharp at the other.

## Calcium Oxalate in Plants

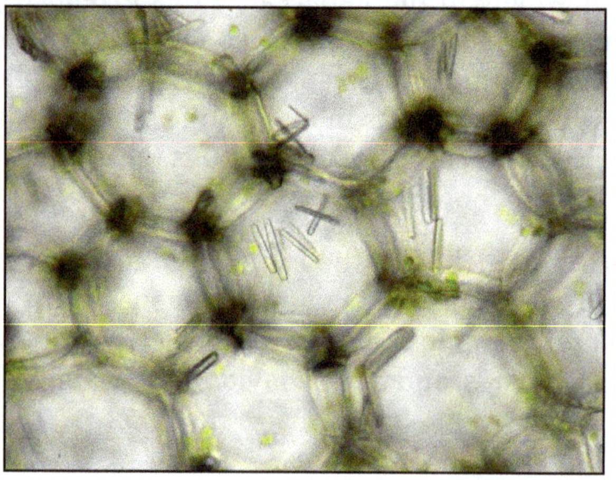

Raphides in *Hypoestes phyllostachya*, the polka dot plant

Many plants accumulate calcium oxalate crystals in response to surplus calcium, which is found throughout the natural environment. The crystals are produced in an intriguing variety of shapes. The crystal morphology depends on the taxonomic group of the plant. In one study of over 100 species, it was found that calcium oxalate accounted for 6.3% of plant dry weight. Crystal morphology and the distribution of raphides (in roots or leaves or tubers etc.) is similar in same taxa but different in others leaving possible opportunities for plant key characteristics and systematic identification; mucilage in raphide containing cells makes light microscopy difficult, though. Little is known about the mechanisms of sequestration or indeed the reason for accumulation of raphides but it is most likely as a defense mechanism against herbivory. It has also been suggested that in some cases raphides may help form plant skeletal structure. Raphides typically occur in parenchyma cells in aerial organs especially the leaves, and are generally confined to the mesophyl. As the leaf area increases, so does the number of raphides, the process starting in even young leaves. The first indications that the cell will contain crystals is shown when the cells enlarge with a larger nucleus.

Raphides are found in specialized plant cells or crystal chambers called idioblasts. Electron micrographs have shown that raphide needle crystals are normally four sided or H-shaped (with a groove down both sides) or with a hexagonal cross section and some are barbed. Wattendorf (1976) suggested that all circular sectioned raphides, as visible in a light microscope, are probably hexagonal in cross section Microscopy using polarized light shows bright opalescence with raphides. Plants like Tradescantia pallida also accumulate calcium oxalate crystals in response to heavy metals stress.

### Harmful Effects

Raphides can produce severe toxic reactions by facilitating the passage of toxin through the herbivore's skin when the tissue containing the raphides also contains toxins. The lethal dose to mice is around 15 mg/kg. Raphides seem to be a defense mechanism against plant predators, as they are likely to tear and harm the soft tissues of the throat or esophagus of a plant predator chewing on the plant's leaves. The venomous process is in two stages: mechanical pricking and injection of harmful protease. Typically ingestion of plants containing raphides, like those common in certain houseplants, can cause immediate numbing followed shortly by painful edema, vesicle formation and dysphagia accompanied by painful stinging and burning to the mouth and throat with symptoms occurring for up to two weeks. Airway assessment and management are of the highest priority, as are extensive irrigation and analgesics in eye exposure.

Raphides cannot normally be destroyed by boiling; that requires an acidic environment or chemical solvents like ether, but heating raphide containing plant materials (like tubers) can fix the raphides into a dried starchy matrix so they are no longer mobile and thus less harmful. Some other plants store raphides in mucilaginous environments and also do not taste acrid.

# Plant Tolerance to Herbivory

Tolerance is the ability of plants to mitigate the negative fitness effects caused by herbivory. It is one of the general plant defense strategies against herbivores, the other being resistance, which is the ability of plants to prevent damage (Strauss and Agrawal 1999). Plant defense strategies play important roles in the survival of plants as they are fed upon by many different types of herbivores, especially insects, which may impose negative fitness effects (Strauss and Zangerl 2002). Damage can occur in almost any part of the plants, including the roots, stems, leaves, flowers and seeds (Strauss and Zergerl 2002). In response to herbivory, plants have evolved a wide variety of defense mechanisms and although relatively less studied than resistance strategies, tolerance traits play a major role in plant defense (Strauss and Zergerl 2002, Rosenthal and Kotanen 1995).

Traits that confer tolerance are controlled genetically and therefore are heritable traits under selection (Strauss and Agrawal 1999). Many factors intrinsic to the plants, such as growth rate, storage capacity, photosynthetic rates and nutrient allocation and uptake, can affect the extent to which plants can tolerate damage (Rosenthal and Kotanen 1994). Extrinsic factors such as soil nutrition, carbon dioxide levels, light levels, water availability and competition also have an effect on tolerance (Rosenthal and Kotanen 1994).

## History of The Study of Plant Tolerance

Studies of tolerance to herbivory has historically been the focus of agricultural scientists (Painter 1958; Bardner and Fletcher 1974). Tolerance was actually initially classified as a form of resistance (Painter 1958). Agricultural studies on tolerance, however, are mainly concerned with the compensatory effect on the plants' yield and not its fitness, since it is of economical interest to reduce crop losses due to herbivory by pests (Trumble 1993; Bardner and Fletcher 1974). One surprising discovery made about plant tolerance was that plants can overcompensate for the damaged caused by herbivory, causing controversy whether herbivores and plants can actually form a mutualistic relationship (Belsky 1986).

It was soon recognized that many factors involved in plants tolerance, such as photosynthetic rates and nutrient allocation, were also traits intrinsic to plant growth and so resource availability may play an important role (Hilbert et al. 1981; Maschinski and Whitham 1989). The growth rate model proposed by Hilbert et al. (1981) predicts plants have higher tolerance in environments that does not allow it to grow at maximum capacity, while the compensatory continuum hypothesis by Maschinski and Whitham (1989) predicts higher tolerance in resource rich environments. Although it was the latter that received higher acceptance, 20 years later, the limiting resource model was proposed to explain the lack of agreement between empirical data and existing models (Wise and Abrahamson 2007). Currently, the limiting resource model is able to explain much more of the empirical data on plant tolerance relative to either of the previous models (Wise and Abrahamson 2008a).

It was only recently that the assumption that tolerance and resistance must be negatively associated has been rejected (Nunez-Farfan et al. 2007). The classical assumption that tolerance traits confer no negative fitness consequences on herbivores has also been questioned (Stinchcombe 2002). Further studies using techniques in quantitative genetics have also provided evidence that tolerance to herbivory is heritable (Fornoni 2011). Studies of plant tolerance have only received increased attention recently, unlike resistance traits which were much more heavily studied (Fornoni 2011). Many aspects of plant tolerance such as its geographic variation, its macroevolutionary implications and its coevolutionary effects on herbivores are still relatively unknown (Fornoni 2011).

## Mechanisms of Tolerance

Plants utilize many mechanisms to recover fitness from damage. Such traits include increased photosynthetic activity, compensatory growth, phenological changes, utilizing stored reserves, reallocating resources, increase in nutrients uptake, and plant architecture (Rosenthal and Kotanen 1994; Strauss and Agrawal 1999; Tiffin 2000).

## Photosynthetic Rates

An increase in photosynthetic rate in undamaged tissues is commonly cited as a mechanism for plants to achieve tolerance (Trumble et al. 1993; Strauss and Agrawal 1999). This is possible since leaves often function at below their maximum capacity (Trumble et al. 1993). Several different pathways may lead to increases in photosynthesis, including higher levels of the Rubisco enzyme and delays in leaf senescence (Stowe et al. 2000). However, detecting an increase in photosynthesis does not mean plants are tolerant to damage. The resources gained from these mechanisms can be used to increase resistance instead of tolerance, such as for the production secondary compounds in the plant (Tiffin 2000). Also, whether the increase in photosynthetic rate is able to compensate for the damage is still not well studied (Trumble et al. 1993; Stowe et al. 2000).

## Compensatory Growth

Biomass regrowth following herbivory is often reported as an indicator of tolerance and plant response after apical meristem damage (AMD) is one of the most heavily studied mechanisms of tolerance (Tiffin 2000; Suwa and Maherali 2008; Wise and Abrahamson 2008). Meristems are sites of rapid cell divisions and so have higher nutrition than most other tissues on the plants . Damage to apical meristems of plants may release it from apical dominance, activating the growth of axillary meristems which increases branching (Trumble et al. 1993; Wise and Abrahamson 2008). Studies have found branching after AMD to undercompensate, fully compensate and overcompensate for the damage received (Marquis 1996, Haukioja and Koricheva 2000, Wise and Abrahamson 2008). The variation in the extent of growth following herbivory may depend on the number and distribution of meristems, the pattern in which they are activated and the number of new meristems

(Stowe et al. 2000). The wide occurrence of overcompensation after AMD has also brought up a controversial idea that there may be a mutualistic relationship between plants and their herbivores (Belsky 1986; Agrawal 2000; Edwards 2009). As will be further discussed below, herbivores may actually be mutualists of plants, such as *Ipomopsis aggregata*, which overcompensate for herbivory (Edwards 2009). Although there are many examples showing biomass regrowth following herbivory, it has been criticized as a useful predictor of fitness since the resources used for regrowth may translate to fewer resources allocated to reproduction (Suwa and Maherali 2008).

## Phenological Change

Studies have shown herbivory can cause delays in plant growth, flowering and fruit production (Tiffin 2000). How plants respond to these phenological delays is likely a tolerance mechanism that will depend highly on their life history and other ecological factors such as, the abundance of pollinators at different times during the season (Tiffin 2000). If the growing season is short, plants that are able to shorten the delay of seed production caused by herbivory are more tolerant than those that cannot shorten this phenological change (Tiffin 2000). These faster recovering plant will be selectively favored over those that cannot as they will pass on more of their offspring to the next generation. In longer growing seasons, however, there may be enough time for most plants to produce seeds before the season ends regardless of damage. In this case, plants that can shorten the phenological delay are not any more tolerant than those that cannot as all plants can reproduce before the season ends (Tiffin 2000).

## Stored Reserves and Resource Reallocation

Resource allocation following herbivory is commonly studied in agricultural systems (Trumble et al. 1993). Resources are most often allocated to reproductive structures after damage, as shown by Irwin et al. (2008) in which *Polemonium viscosum* and *Ipomopsis aggregata* increased flower production after flower larceny. When these reproductive structures are not present, resources are allocated to other tissues, such as leaves and shoots as seen in juvenile *Plantago lanceolata* (Trumble et al. 1993; Barton 2008). Utilizing stored reserves may be an important tolerance mechanism for plants which have abundant time to collect and store resources, such as perennial plants (Tiffin 2000; Erb *et al.* 2009). Resources are often stored in leaves and specialized storage organs such as tubers and roots, and studies have shown evidence that these resources are allocated for regrowth following herbivory (Trumble et al. 1993; Tiffin 2000; Erb *et al.* 2009). However, the importance of this mechanism to tolerance is not well studied and it is unknown how much it contributes to tolerance since stored reserves mostly consist of carbon resources, whereas tissue damage causes a loss of carbon, nitrogen and other nutrients (Tiffin 2000).

## Plant Architecture

This form of tolerance relies on constitutive mechanisms, such as morphology, at the time of damage, unlike the induced mechanisms mentioned above. Plant architecture includes roots to shoots ratios, stem number, stem rigidity and plant vasculature (Marquis 1996, Tiffin 2000). A high roots to shoots ratio will allow plants to better absorb nutrients following herbivory and rigid stems will prevent collapse after sustaining damage, increasing plant tolerance (Tiffin 2000). Since plants have a meristemic construction, how resources are restricted among different regions of the plants, referred to as sectoriality, will affect the ability to transfer resources from undamaged areas to damaged areas (Marquis

1996). Although plant vasculature may play important roles in tolerance, it is not well studied due to the difficulties in identifying the flow of resources (Marquis 1996). Increasing a plant's vasculature would seem advantageous since it increases the flow of resources to all sites of damage but it may also increase its susceptibility to herbivores, such as phloem suckers (Marquis 1996, Stowe et al. 2000).

## Measuring Tolerance

Tolerance is operationally defined as the slope of the regression between fitness and level of damage (Stinchcombe 2002). Since an individual plant can only sustain one level of damage, it is necessary to measure fitness using a group of related individuals, preferably full-sibs or clones to minimize other factors that may influence tolerance, after sustaining different levels of damage (Stinchcombe 2002). Tolerance is often presented as a reaction norm, where slopes larger than, equal to and less than zero reflect overcompensation, full compensation and undercompensation, respectively (Strauss and Agrawal 1999).

## Scales of Measurement

Both fitness and herbivory can be measured or analyzed using an absolute (additive) scale or a relative (multiplicative) scale (Wise and Carr 2008b). The absolute scale may refer to number of fruits produced or total area of leaf eaten, while the relative scale may refer to proportion of fruits damaged or proportion of leaves eaten. Wise and Carr (2008b) suggested that it is best to keep the measure of fitness and the measure of damage on the same scale when analyzing tolerance since having them on different scales may result is misleading outcomes. Even if the data were measured using different scales, data on the absolute scale can be log-transformed to be more similar to data on a relative (multiplicative) scale (Wise and Carr 2008b).

## Simulated Vs Natural Herbivory

A majority of studies use simulated or manipulated herbivory, such as clipping leaves or herbivore exclusions, due to the difficulty in controlling damage levels under natural conditions (Tiffin and Inouye 2000). The advantage of using natural herbivory is that plants will experience the pattern of damage that selection has favored tolerance for, but there may be biases resulting from unmeasured environmental variables that may affect both plant and herbivores. Using simulated herbivory allows for the control of environmental variables, but replicating natural herbivory is difficult, causing plants to respond differently from imposed and natural herbivory (Tiffin and Inouye 2000). Growing plants in the control environment of the greenhouse may also affect their response as it is still a novel environment to the plants. Even if the plots are grown in natural settings, the methods of excluding or including herbivores, such as using cages or pesticides, may also affect plant tolerance (Tiffin and Inouye 2000). Lastly, models have predicted that manipulated herbivory may actually result in less precise estimates of tolerance relative to that from natural herbivory (Tiffin and Inoue 2000).

## Fitness Traits

Many studies have shown that using different measurements of fitness may give varying outcomes of tolerance (Strauss and Agrawal 1999; Suwa and Maherali 2008; Banta et al. 2010). Banta et al.

(2010) found that their measure of tolerance will differ depending on whether fruit production or total viable seed production was used to reflect fitness in *Arabdopsis thaliana*. Careful considerations must be made to choose traits that are linked to fitness as closely as possible when measuring tolerance.

## Tolerance-resistance Trade-off

It is classically assumed that there is a negative correlation between the levels of tolerance and resistance in plants (Stowe et al. 2000; Nunez-Farfan et al. 2007). For this trade-off to exist, it requires that tolerance and resistance be redundant defense strategies with similar costs to the plant (Nunez-Farfan et al. 2007). If this is the case, then plants that are able to tolerate damage will suffer little decrease in fitness and so resistance would not be selectively favored. For highly resistant plants, allocating resources to tolerance would not be selectively favored as the plant received minimal damage in the first place.

There is now increasing evidence that many plants allocate resources to both types of defense strategies (Nunez-Farfan et al. 2007). There is also evidence that there may not be a trade-off between tolerance and resistance at all and that they may evolve independently (Leimu and Koricheva 2006; Nunez-Farfan et al. 2007; Muola et al. 2010). Models have shown that intermediate levels of resistance and tolerance are evolutionary stable as long as the benefits of having both traits are more than additive (Nunez-Farfan et al. 2007). Tolerance and resistance may not be redundant strategies since tolerance could be necessary for damage from large mammalian herbivores or specialist herbivores which have the ability to circumvent resistance traits of the plant (Nunez-Farfan et al. 2007; Muola et al. 2010). Also, as traits that confer tolerance are usually basic characteristics of plants, the result of selection on growth and not herbivory may also affect tolerance (Rosenthal and Kotanen 1994).

## Ontogenetic Shifts

It has been suggested that the trade-off between resistance and tolerance may change throughout the development of the plants. It is often assumed that seedlings and juveniles are less tolerant of herbivory since they did not develop the structures required for resource acquisition and so will rely more on traits that confer resistance (Boege et al. 2007; Barton 2008, Barton and Koricheva 2010; Tucker and Avila-Sakar 2010). Although many studies find lower tolerance in seedlings, this is not always the case, as seen in juveniles of *Plant ago lanceolata* which can fully compensate for 50% defoliation (Barton 2008). There is also the added complexity of shifts in herbivore communities as the plant develops and so may favor tolerance or resistance at different life stages (Barton and Koricheva 2010).

## Effects of Resource Levels on Tolerance

The response of plants to herbivory is often plastic and varies according to the conditions it is experiencing (Wise and Abrahamson 2005). The major resources that affect plant growth and also tolerance are water, light, carbon dioxide and soil nutrients. Water and light levels are generally assumed to be positively associated with tolerance (Strauss and Agrawal 1999). However, there are exceptions such as evidence of decreased tolerance in *Madia sativa* with increased water availability (Wise and Abrahamson 2007, Gonzales *et al.* 2008). Many studies have found $CO_2$ levels to decrease tolerance in plants (Lau and Tiffin 2009). Increased nutrient levels are also commonly found to be negatively associated with tolerance (Wise and Abrahamson 2007).

There are currently three prominent models that predict how resource levels may alter a plants 's tolerance to herbivory.

## Growth Rate Model (GRM)

The GRM proposes that the growth rate of the plant at the time of damage is important in determining its response (Hilbert et al. 1981). Plants that are growing in stressful conditions, such as low resource levels or high competition, are growing below their maximum growth rate and so may have a higher capacity for regrowth after receiving damage (Hilbert et al. 1981). In contrast, plants in relatively benign conditions are growing near their maximum growth rate. These plants are less able to recover from damage since they are already near their innate maximum growth rate (Hilbert et al. 1981).

## Compensatory Continuum Hypothesis (CCH)

The CCH suggests that there is a continuum of responses to herbivory (Maschinski and Whitham 1989). It predicts that plants growing in less stressful environment conditions, such as high resource or low competition, are better able to tolerate herbivory since they have abundant resources to replace lost tissues and recover from the damage. Plants growing in stressful environments are then predicted to have lower tolerance (Maschinski and Whitham 1989).

## Limiting Resource Model (LRM)

This recently proposed model takes into account the resource that is limiting plant fitness, the resource affected by herbivory and how the acquisition of resources is affected by herbivory (Wise and Abrahamson 2005). Unlike the GRM and CCH, it is able to incorporate the type of damage received since different modes of herbivory may cause different resources to be affected by herbivory. The LRM encompasses every possible outcome of tolerance (i.e. equal tolerance in both environments, higher tolerance in low stress environments and lower tolerance in low stress environments) and allows multiple pathways to reach these outcome.

Currently, the LRM seems to be most useful in predicting the effects that varying resources levels may have on tolerance (Wise and Abrahamson 2007). Meta-analyses by Hawkes and Sullivan (2001) and Wise and Abrahamson (2007, 2008a) found that the CCH and GRM were insufficient in predicting the diversity of plant tolerance to herbivory. Banta et al. (2010), however, suggested that the LRM should be represented as a set of seven models, instead of one, since each individual part of the LRM requires different assumptions.

## Selection on Herbivores

It is classically assumed that tolerance traits do not impose selection on herbivore fitness (Strauss and Agrawal 1999). This is in contrast to traits that confer resistance, which are likely to affect herbivore fitness and lead to a co-evolutionary arms race (Stinchcombe 2002; Espinosa and Fornoni 2006). However, there are possible mechanisms in which tolerance may affect herbivore fitness .

One mechanism requires a genetic association between loci that confers resistance and tolerance either through tight linkage or pleiotropy (Stinchcombe 2002). Selection for either trait will then also affect the other. If there is a positive correlation between the two traits, then selection for increased tolerance will also increase resistance in the plants. If there is a negative correlation between the two traits then selection for increased tolerance will decrease resistance. How common this association exists, however, is uncertain as there are many studies which find no correlation between tolerance and resistance and others which find significant correlations between them (Leimu and Koricheva 2006; Nunez-Farfan et al. 2007; Muola et al. 2010).

If the traits that allow for tolerance affects the plant tissue's quality, quantity or availability, tolerance may also impose selection on herbivores. Consider a case where tolerance is achieved through activation of dormant meristems in the plants . These new plant tissues may be of lower quality than what was previously eaten by herbivores. herbivores which have higher rates of consumption or can more efficiently use this new resource may be selectively favored over those that cannot (Stinchcombe 2002).

Espinosa and Fornoni (2006) was one study which directly investigated whether tolerance may impose selection on herbivores. As suggested by Stinchcombe (2002), they used plants which had similar resistance but differed in tolerance to more easily differentiate the effects of each trait. As expected, they found evidence that resistance in plants affected herbivore fitness, but they were unable to find any effects of tolerance on herbivore fitness .

A recent model by Restif and Koella (2003) found that plant tolerance can directly impose selection on pathogens. Assuming that investment in tolerance will reduce plant fecundity, infection by pathogens will decrease the number of uninfected hosts. There may then be selection for decreased virulence in the pathogens, so that their plant host will survive long enough to produce enough offspring for future pathogens to infect (Restif and Koelle 2003). However, this may only have limited application to herbivores.

## Species Interactions

### Plant Communities

Herbivory can have large effects on the succession and diversity of plants communities (Anderson and Briske 1995; Stowe et al. 2000; Pejman et al. 2009). Thus, plant defense strategies are important in determining temporal and spatial variation of plant species as it may change the competitive abilities of plants following herbivory. (Anderson and Briske 1995; Stowe et al. 2000).

Past studies have suggested plant resistance to play the major role in species diversity within communities, but tolerance may also be an important factor (Stowe et al. 2000; Pejman et al. 2009). Herbivory may allow less competitive, but tolerant plants to survive in communities dominated by highly competitive but intolerant plant species, thereby increasing diversity (Mariotte et al. 2013). Pejman et al. (2009) found support for this idea in an experimental study on grassland species. In low resource environments, highly competitive (dominant) plants species had lower tolerance than the less competitive (subordinate) species. They also found that the addition of fertilizers offset the negative effects of herbivory on dominant plants. It has also been suggested that the observation of species that occur late in ecological succession (late-seral) being replaced by species that occur in the middle of ecological succession (mid-seral) after high herbivory is due to differences in tolerance between them (Anderson and Briske 1995; Off and Ritchie 1998). However, tolerance between these two groups of species do not always differ and other factors, such as selective herbivory on late-seral species, may contribute to these observations (Anderson and Briske 1995).

## Mutualisms

The large number of studies indicating overcompensation in plants following herbivory, especially after apical meristem damage, has led some authors to suggest that there may be mutualistic relationships between plants and herbivores (Belsky 1986; Agrawal 2000; Edwards 2009). If herbivores provide some benefit for the plant despite causing damage, the plant may evolve tolerance to minimize the damage imposed by the herbivore to shift the relationship more towards mutualism (Edwards 2009). Such benefits include the release from apical dominance, inducing resistance traits to temporally separate herbivores, providing information of future attacks and pollination (Agrawal 2000).

Ipomopsis aggregata

One of the best examples occurs in *Ipomopsis aggregata* where there is increased seed production and seed siring in damaged plants compared to undamaged plants (Figure 4; Edwards 2009). The probability of attack after the first bout of herbivory is low in the environment inhabited by *I. aggregata*. Due to the predictability of attacks, these plants have evolved to overcompensate for the damage and produce the majority of their seeds after the initial bout of herbivory (Edwards 2009). Another example involves endophytic fungi, such as *Neophtodium*, which parasitize plants and produce spores that destroy host inflorescences (Edwards 2009). The fungi also produce alkaloids which protect the plant from herbivores and so the plant may have evolved tolerance to flower damage to acquire this benefit (Edwards 2009). Tolerance may also be involved in the mutualism between the myremecophyte, *Cordia nodosa*, and its ant symbiont *Allomerus octoarticulatus* (Edwards and Yu, 2008). The plant provides the ant with shelter and food bodies in return for protection against herbivory, but the ants also sterilize the plant by removing flower buds. *C. nodosa* is able to compensate for this by reallocating resources to produce flowers on branches not occupied by castrating ants (Edwards and Yu, 2008).

A similar type of mutualism involves plants and mycorrhizal fungi (Bennett and Bever 2007). Mycorrhizal fungi inhabit plant roots and increase nutrient uptake for the plant in exchange for food resources. These fungi are also able to alter the tolerance of plants to herbivory and may cause undercompesation, full compensation and overcompensation depending of the species of fungi involved (Bennett and Bever 2007).

## Agriculture

Modern agriculture has focuses on using genetically modified crops which possess toxic compounds to reduce damage by pests (Nunez-Farfan et al. 2007). However, the effectiveness of resistance traits may decrease as herbivores evolve counter adaptations to the toxic compound, especially since most farmers are reluctant to assign a proportion of their land to contain susceptible crops (Nunez-Farfan et al. 2007). Another method to increase crop yield is to use lines that are tolerant to herbivory and can compensate or even overcompensate for the damage inflicted (Nunez-Farfan et al. 2007; Poveda et al. 2010).

Alterations in resource allocation due to herbivory is studied heavily in agricultural systems (Trumble et al. 1993). Domestication of plants by selecting for higher yield have undoubtedly also caused changes in various plant growth traits, such as decreased resource allocation to non-yield tissues (Welter and Steggall 1993). Alterations in growth traits is likely to effect plant tolerance since the mechanisms overlap. That domesticated tomato plants have lower tolerance to folivory than their wild progenitors suggests this as well (Welter and Steggall 1993). Most agricultural studies however, are more focused on comparing tolerance between damaged and undamaged crops, not between crops and their wild counterparts. Many have found crops, such as cucumbers, cabbages and cauliflowers, can fully compensate and overcompensate for the damaged received (Trumble et al. 1993). A recent study by Poveda et al. (2010) also found evidence of overcompensation in potato plants in response to tuber damage by the potato tuber moth, *Phthorimaea operculella*. Unlike previous examples, the potato plant does not reallocate resources, but actually increases overall productivity to increase mass of tubers and aboveground tissues (Poveda et al. 2010).

# Plant Defense Against Herbivory

Foxgloves produce several deadly chemicals, namely cardiac and steroidal glycosides. Ingestion can cause nausea, vomiting, hallucinations, convulsions, or death.

Plant defense against herbivory or host-plant resistance (HPR) describes a range of adaptations evolved by plants which improve their survival and reproduction by reducing the impact of herbivores. Plants can sense being touched, and they can use several strategies to defend against damage caused by herbivores. Many plants produce secondary metabolites, known as allelochemicals, that influence the behavior, growth, or survival of herbivores. These chemical defenses can act as repellents or toxins to herbivores, or reduce plant digestibility.

Other defensive strategies used by plants include escaping or avoiding herbivores in any time and/or any place, for example by growing in a location where plants are not easily found or accessed by herbivores, or by changing seasonal growth patterns. Another approach diverts herbivores toward eating non-essential parts, or enhances the ability of a plant to recover from the damage caused by herbivory. Some plants encourage the presence of natural enemies of herbivores, which in turn protect the plant. Each type of defense can be either *constitutive* (always present in the plant), or *induced* (produced in reaction to damage or stress caused by herbivores).

Historically, insects have been the most significant herbivores, and the evolution of land plants is closely associated with the evolution of insects. While most plant defenses are directed against insects, other defenses have evolved that are aimed at vertebrate herbivores, such as birds and mammals. The study of plant defenses against herbivory is important, not only from an evolutionary view point, but also in the direct impact that these defenses have on agriculture, including human and livestock food sources; as beneficial 'biological control agents' in biological pest control programs; as well as in the search for plants of medical importance.

## Evolution of Defensive Traits

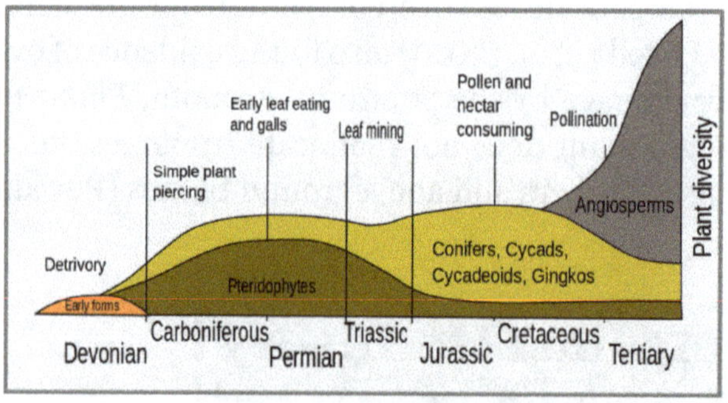


Timeline of plant evolution and the beginnings of different modes of insect herbivory

The earliest land plants evolved from aquatic plants around 450 million years ago (Ma) in the Ordovician period. Many plants have adapted to iodine-deficient terrestrial environment by removing iodine from their metabolism, in fact iodine is essential only for animal cells. An important antiparasitic action is caused by the block of the transport of iodide of animal cells inhibiting sodium-iodide symporter (NIS). Many plant pesticides are glycosides (as the cardiac digitoxin) and cyanogenic glycosides which liberate cyanide, which, blocking cytochrome c oxidase and NIS, is poisonous only for a large part of parasites and herbivores and not for the plant cells in which it seems useful in seed dormancy phase. Iodide is not pesticide, but is oxidized, by vegetable peroxidase, to iodine, which is a strong oxidant, it is able to kill bacteria, fungi and protozoa.

The Cretaceous period saw the appearance of more plant defense mechanisms. The diversification of flowering plants (angiosperms) at that time is associated with the sudden burst of speciation in insects.

This diversification of insects represented a major selective force in plant evolution, and led to selection of plants that had defensive adaptations. Early insect herbivores were mandibulate and bit or chewed vegetation; but the evolution of vascular plants lead to the co-evolution of other forms of herbivory, such as sap-sucking, leaf mining, gall forming and nectar-feeding.

The relative abundance of different species of plants in ecological communities including forests and grasslands may be determined in part by the level of defensive compounds in the different species. Since the cost of replacement of damaged leaves is higher in conditions where resources are scarce, it may also be that plants growing in areas where water and nutrients are scarce may invest more resources into anti-herbivore defenses.

## Records of Herbivores

*Viburnum lesquereuxii* leaf with insect damage; Dakota Sandstone (Cretaceous) of Ellsworth County, Kansas. Scale bar is 10 mm.

Our understanding of herbivory in geological time comes from three sources: fossilized plants, which may preserve evidence of defense (such as spines), or herbivory-related damage; the observation of plant debris in fossilised animal faeces; and the construction of herbivore mouthparts.

Long thought to be a Mesozoic phenomenon, evidence for herbivory is found almost as soon as fossils which could show it. Within under 20 million years of the first fossils of sporangia and stems towards the close of the Silurian, around 420 million years ago, there is evidence that they were being consumed. Animals fed on the spores of early Devonian plants, and the Rhynie chert also provides evidence that organisms fed on plants using a "pierce and suck" technique. Many plants of this time are preserved with spine-like enations, which may have performed a defensive role before being co-opted to develop into leaves.

During the ensuing 75 million years, plants evolved a range of more complex organs - from roots to seeds. There was a gap of 50 to 100 million years between each organ evolving, and it being fed upon. Hole feeding and skeletonization are recorded in the early Permian, with surface fluid feeding evolving by the end of that period.

A Plain Tiger *Danaus chrysippus* caterpillar making a moat to block defensive chemicals of *Calotropis* before feeding

## Co-evolution

Herbivores are dependent on plants for food, and have evolved mechanisms to obtain this food despite the evolution of a diverse arsenal of plant defenses. Herbivore adaptations to plant defense have been likened to *offensive traits* and consist of adaptations that allow increased feeding and use of a host plant. Relationships between herbivores and their host plants often results in reciprocal evolutionary change, called co-evolution. When an herbivore eats a plant it selects for plants that can mount a defensive response. In cases where this relationship demonstrates *specificity* (the evolution of each trait is due to the other), and *reciprocity* (both traits must evolve), the species are thought to have co-evolved. The "escape and radiation" mechanism for co-evolution presents the idea that adaptations in herbivores and their host plants have been the driving force behind speciation, and have played a role in the radiation of insect species during the age of angiosperms. Some herbivores have evolved ways to hijack plant defenses to their own benefit, by sequestering these chemicals and using them to protect themselves from predators. Plant defenses against herbivores are generally not complete so plants also tend to evolve some tolerance to herbivory.

## Types

Plant defenses can be classified generally as constitutive or induced. Constitutive defenses are always present in the plant, while induced defenses are produced or mobilized to the site where a plant is injured. There is wide variation in the composition and concentration of constitutive defenses and these range from mechanical defenses to digestibility reducers and toxins. Many external mechanical defenses and large quantitative defenses are constitutive, as they require large amounts of resources to produce and are difficult to mobilize. A variety of molecular and biochemical approaches are used to determine the mechanism of constitutive and induced plant defenses responses against herbivory.

Induced defenses include secondary metabolic products, as well as morphological and physiological changes. An advantage of inducible, as opposed to constitutive defenses, is that they are only produced when needed, and are therefore potentially less costly, especially when herbivory is variable.

## Chemical Defenses

Persimmon, genus *Diospyros*, has a high tannin content which gives immature fruit, seen above, an astringent and bitter flavor.

The evolution of chemical defences in plants is linked to the emergence of chemical substances that are not involved in the essential photosynthetic and metabolic activities. These substances, secondary metabolites, are organic compounds that are not directly involved in the normal growth, development or reproduction of organisms, and often produced as by-products during the synthesis of primary metabolic products. Although these secondary metabolites have been thought to play a major role in defenses against herbivores, a meta-analysis of recent relevant studies has suggested that they have either a more minimal (when compared to other non-secondary metabolites, such as primary chemistry and physiology) or more complex involvement in defense.

Secondary metabolites are often characterized as either *qualitative* or *quantitative*. Qualitative metabolites are defined as toxins that interfere with an herbivore's metabolism, often by blocking specific biochemical reactions. Qualitative chemicals are present in plants in relatively low concentrations (often less than 2% dry weight), and are not dosage dependent. They are usually small, water-soluble molecules, and therefore can be rapidly synthesized, transported and stored with relatively little energy cost to the plant. Qualitative allelochemicals are usually effective against non-adapted specialists and generalist herbivores.

Quantitative chemicals are those that are present in high concentration in plants (5 – 40% dry weight) and are equally effective against all specialists and generalist herbivores. Most quantitative metabolites are digestibility reducers that make plant cell walls indigestible to animals. The effects of quantitative metabolites are dosage dependent and the higher these chemicals' proportion in the herbivore's diet, the less nutrition the herbivore can gain from ingesting plant tissues. Because they are typically large molecules, these defenses are energetically expensive to produce and maintain, and often take longer to synthesize and transport.

The geranium, for example, produces a unique chemical compound in its petals to defend itself from Japanese beetles. Within 30 minutes of ingestion the chemical paralyzes the herbivore. While the chemical usually wears off within a few hours, during this time the beetle is often consumed by its own predators.

- See Toxalbumin

## Types of Chemical Defences

Plants have evolved many secondary metabolites involved in plant defense, which are collectively known as antiherbivory compounds and can be classified into three sub-groups: nitrogen compounds (including *alkaloids, cyanogenic glycosides, glucosinolates* and *benzoxazinoids*), terpenoids, and phenolics.

Alkaloids are derived from various amino acids. Over 3000 known alkaloids exist, examples include nicotine, caffeine, morphine, cocaine, colchicine, ergolines, strychnine, and quinine. Alkaloids have pharmacological effects on humans and other animals. Some alkaloids can inhibit or activate enzymes, or alter carbohydrate and fat storage by inhibiting the formation phosphodiester bonds involved in their breakdown. Certain alkaloids bind to nucleic acids and can inhibit synthesis of proteins and affect DNA repair mechanisms. Alkaloids can also affect cell membrane and cytoskeletal structure causing the cells to weaken, collapse, or leak, and can affect nerve transmission. Although alkaloids act on a diversity of metabolic systems in humans and other animals, they almost uniformly invoke an aversively bitter taste.

Cyanogenic glycosides are stored in inactive forms in plant vacuoles. They become toxic when herbivores eat the plant and break cell membranes allowing the glycosides to come into contact with enzymes in the cytoplasm releasing hydrogen cyanide which blocks cellular respiration. Glucosinolates are activated in much the same way as cyanogenic glucosides, and the products can cause gastroenteritis, salivation, diarrhea, and irritation of the mouth. Benzoxazinoids, secondary defence metabolites, which are characteristic for grasses (Poaceae), are also stored as inactive glucosides in the plant vacuole. Upon tissue disruption they get into contact with β-glucosidases from the chloroplasts, which enzymatically release the toxic aglucones. Whereas some benzoxazinoids are constitutively present, others are only synthesised following herbivore infestation, and thus, considered inducible plant defenses against herbivory.

The terpenoids, sometimes referred to as isoprenoids, are organic chemicals similar to terpenes, derived from five-carbon isoprene units. There are over 10,000 known types of terpenoids. Most are multicyclic structures which differ from one another in both functional groups, and in basic carbon skeletons. Monoterpenoids, continuing 2 isoprene units, are volatile essential oils such as citronella, limonene, menthol, camphor, and pinene. Diterpenoids, 4 isoprene units, are widely distributed in latex and resins, and can be quite toxic. Diterpenes are responsible for making *Rhododendron* leaves poisonous. Plant steroids and sterols are also produced from terpenoid precursors, including vitamin D, glycosides (such as digitalis) and saponins (which lyse red blood cells of herbivores).

Phenolics, sometimes called *phenols*, consist of an aromatic 6-carbon ring bonded to a hydroxy group. Some phenols have antiseptic properties, while others disrupt endocrine activity. Phenolics range from simple tannins to the more complex flavonoids that give plants much of their red, blue, yellow, and white pigments. Complex phenolics called polyphenols are capable of producing many different types of effects on humans, including antioxidant properties. Some examples of phenolics used for defense in plants are: lignin, silymarin and cannabinoids. Condensed tannins, polymers composed of 2 to 50 (or more) flavonoid molecules, inhibit herbivore digestion by binding to consumed plant proteins and making them more difficult for animals to digest, and by interfering with protein absorption and digestive enzymes. Silica and lignins, which are completely indigestible to animals, grind down insect mandibles (appendages necessary for feeding).

In addition to the three larger groups of substances mentioned above, fatty acid derivates, amino acids and even peptides are also used as defense. The cholinergic toxine, cicutoxin of water hemlock, is a polyyne derived from the fatty acid metabolism. β-N-Oxalyl-L-α,β-diaminopropionic acid as simple amino acid is used by the sweet pea which leads also to intoxication in humans. The synthesis of fluoroacetate in several plants is an example of the use of small molecules to disrupt the metabolism of herbivores, in this case the citric acid cycle.

In tropical *Sargassum* and *Turbinaria* species that are often preferentially consumed by herbivorous fishes and echinoids, there is a relatively low level of phenolics and tannins.

## Mechanical Defenses

The thorns on the stem of this raspberry plant, serve as a mechanical defense against herbivory.

Plants have many external structural defenses that discourage herbivory. Depending on the herbivore's physical characteristics (i.e. size and defensive armor), plant structural defenses on stems and leaves can deter, injure, or kill the grazer. Some defensive compounds are produced internally but are released onto the plant's surface; for example, resins, lignins, silica, and wax cover the epidermis of terrestrial plants and alter the texture of the plant tissue. The leaves of holly plants, for instance, are very smooth and slippery making feeding difficult. Some plants produce gummosis or sap that traps insects.

A plant's leaves and stem may be covered with sharp prickles, spines, thorns, or trichomes- hairs on the leaf often with barbs, sometimes containing irritants or poisons. Plant structural features like spines and thorns reduce feeding by large ungulate herbivores (e.g. kudu, impala, and goats) by restricting the herbivores' feeding rate, or by wearing down the molars. Raphides are sharp needles of calcium oxalate or calcium carbonate in plant tissues, making ingestion painful, damaging a herbivore's mouth and gullet and causing more efficient delivery of the plant's toxins. The structure of a plant, its branching and leaf arrangement may also be evolved to reduce herbivore impact. The shrubs of New Zealand have evolved special wide branching adaptations believed to be a response to browsing birds such as the moas. Similarly, African Acacias have long spines low in the canopy, but very short spines high in the canopy, which is comparatively safe from herbivores such as giraffes.

Coconut palms protect their fruit by surrounding it with multiple layers of armor.

Trees such as coconut and other palms, may protect their fruit by multiple layers of armor, needing efficient tools to break through to the seed contents, and special skills to climb the tall and relatively smooth trunk.

## Thigmonasty

Thigmonastic movements, those that occur in response to touch, are used as a defense in some plants. The leaves of the sensitive plant, *Mimosa pudica*, close up rapidly in response to direct touch, vibration, or even electrical and thermal stimuli. The proximate cause of this mechanical response is an abrupt change in the turgor pressure in the pulvini at the base of leaves resulting from osmotic phenomena. This is then spread via both electrical and chemical means through the plant; only a single leaflet need be disturbed.

This response lowers the surface area available to herbivores, which are presented with the underside of each leaflet, and results in a wilted appearance. It may also physically dislodge small herbivores, such as insects.

## Mimicry and Camouflage

Some plants mimic the presence of insect eggs on their leaves, dissuading insect species from laying their eggs there. Because female butterflies are less likely to lay their eggs on plants that already have butterfly eggs, some species of neotropical vines of the genus *Passiflora* (Passion flowers) contain physical structures resembling the yellow eggs of *Heliconius* butterflies on their leaves, which discourage oviposition by butterflies.

## Indirect Defenses

The large thorn-like stipules of *Acacia collinsii* are hollow and offer shelter for ants, which in return protect the plant against herbivores.

Another category of plant defenses are those features that indirectly protect the plant by enhancing the probability of attracting the natural enemies of herbivores. Such an arrangement is known as mutualism, in this case of the "enemy of my enemy" variety. One such feature are semiochemicals, given off by plants. Semiochemicals are a group of volatile organic compounds involved in interactions between organisms. One group of semiochemicals are allelochemicals; consisting of allomones, which play a defensive role in interspecies communication, and kairomones, which are used by members of higher trophic levels to locate food sources. When a plant is attacked it releases allelochemics containing an abnormal ratio of these herbivore-induced plant volatiles (HIPVs). Predators sense these volatiles as food cues, attracting them to the damaged plant, and to feeding herbivores. The subsequent reduction in the number of herbivores confers a fitness benefit to the plant and demonstrates the indirect defensive capabilities of semiochemicals. Induced volatiles also have drawbacks, however; some studies have suggested that these volatiles also attract herbivores.

Plants also provide housing and food items for natural enemies of herbivores, known as "biotic" defense mechanisms, as a means to maintain their presence. For example, trees from the genus *Macaranga* have adapted their thin stem walls to create ideal housing for an ant species (genus

*Crematogaster*), which, in turn, protects the plant from herbivores. In addition to providing housing, the plant also provides the ant with its exclusive food source; from the food bodies produced by the plant. Similarly, some *Acacia* tree species have developed thorns that are swollen at the base, forming a hollowing structure that acts as housing. These *Acacia* trees also produce nectar in extrafloral nectaries on their leaves as food for the ants.

Plant use of endophytic fungi in defense is a very common phenomenon. Most plants have endophytes, microbial organisms that live within them. While some cause disease, others protect plants from herbivores and pathogenic microbes. Endophytes can help the plant by producing toxins harmful to other organisms that would attack the plant, such as alkaloid producing fungi which are common in grasses such as tall fescue (*Festuca arundinacea*).

## Leaf Shedding and Color

There have been suggestions that leaf shedding may be a response that provides protection against diseases and certain kinds of pests such as leaf miners and gall forming insects. Other responses such as the change of leaf colors prior to fall have also been suggested as adaptations that may help undermine the camouflage of herbivores. Autumn leaf color has also been suggested to act as an honest warning signal of defensive commitment towards insect pests that migrate to the trees in autumn.

## Costs and Benefits

Defensive structures and chemicals are costly as they require resources that could otherwise be used by plants to maximize growth and reproduction. Many models have been proposed to explore how and why some plants make this investment in defenses against herbivores.

## Optimal Defense Hypothesis

The optimal defense hypothesis attempts to explain how the kinds of defenses a particular plant might use reflect the threats each individual plant faces. This model considers three main factors, namely: risk of attack, value of the plant part, and the cost of defense.

The first factor determining optimal defense is risk: how likely is it that a plant or certain plant parts will be attacked? This is also related to the *plant apparency hypothesis*, which states that a plant will invest heavily in broadly effective defenses when the plant is easily found by herbivores. Examples of apparent plants that produce generalized protections include long-living trees, shrubs, and perennial grasses. Unapparent plants, such as short-lived plants of early successional stages, on the other hand, preferentially invest in small amounts of qualitative toxins that are effective against all but the most specialized herbivores.

The second factor is the value of protection: would the plant be less able to survive and reproduce after removal of part of its structure by a herbivore? Not all plant parts are of equal evolutionary value, thus valuable parts contain more defenses. A plant's stage of development at the time of feeding also affects the resulting change in fitness. Experimentally, the fitness value of a plant structure is determined by removing that part of the plant and observing the effect. In general, reproductive parts are not as easily replaced as vegetative parts, terminal leaves have greater value than basal leaves, and the loss of plant parts mid-season has a greater negative effect on fitness than removal at the beginning or end of the season. Seeds in particular tend to be very well protected. For example, the seeds of many edible fruits and nuts contain cyanogenic glycosides such as amygdalin. This results from the need to balance the effort needed to make the fruit attractive to animal dispersers while ensuring that the seeds are not destroyed by the animal.

The final consideration is cost: how much will a particular defensive strategy cost a plant in energy and materials? This is particularly important, as energy spent on defense cannot be used for other functions, such as reproduction and growth. The optimal defense hypothesis predicts that plants will allocate more energy towards defense when the benefits of protection outweigh the costs, specifically in situations where there is high herbivore pressure.

## Carbon:nutrient Balance Hypothesis

The carbon:nutrient balance hypothesis, also known as the *environmental constraint hypothesis* or *Carbon Nutrient Balance Model* (CNBM), states that the various types of plant defenses are responses to variations in the levels of nutrients in the environment. This hypothesis predicts the Carbon/Nitrogen ratio in plants determines which secondary metabolites will be synthesized. For example, plants growing in nitrogen-poor soils will use carbon-based defenses (mostly digestibility reducers), while those growing in low-carbon environments (such as shady conditions) are more likely to produce nitrogen-based toxins. The hypothesis further predicts that plants can change their defenses in response to changes in nutrients. For example, if plants are grown in low-nitrogen conditions, then these plants will implement a defensive strategy composed of constitutive carbon-based defenses. If nutrient levels subsequently increase, by for example the addition of fertilizers, these carbon-based defenses will decrease.

## Growth Rate Hypothesis

The growth rate hypothesis, also known as the *resource availability hypothesis*, states that defense strategies are determined by the inherent growth rate of the plant, which is in turn determined by the resources available to the plant. A major assumption is that available resources are the limiting factor in determining the maximum growth rate of a plant species. This model predicts that the level of defense investment will increase as the potential of growth decreases. Additionally, plants in resource-poor areas, with inherently slow-growth rates, tend to have long-lived leaves and twigs, and the loss of plant appendages may result in a loss of scarce and valuable nutrients.

A recent test of this model involved a reciprocal transplants of seedlings of 20 species of trees between clay soils (nutrient rich) and white sand (nutrient poor) to determine whether trade-offs between growth rate and defenses restrict species to one habitat. When planted in white sand and protected from herbivores, seedlings originating from clay outgrew those originating from the nutrient-poor sand, but in the presence of herbivores the seedlings originating from white sand performed better, likely due to their higher levels of constitutive carbon-based defenses. These finding suggest that defensive strategies limit the habitats of some plants.

## Growth-differentiation Balance Hypothesis

The growth-differentiation balance hypothesis states that plant defenses are a result of a tradeoff between "growth-related processes" and "differentiation-related processes" in different environments. Differentiation-related processes are defined as "processes that enhance the structure or function of existing cells (i.e. maturation and specialization)." A plant will produce chemical defenses only when energy is available from photosynthesis, and plants with the highest concentrations of secondary metabolites are the ones with an intermediate level of available resources. The GDBH also accounts for tradeoffs between growth and defense over a resource availability gradient. In situations where resources (e.g. water and nutrients) limit photosynthesis, carbon supply is predicted to limit both growth and defense. As resource availability increases, the requirements needed to support photosynthesis are met, allowing for accumulation of carbohydrate in tissues. As resources are not sufficient to meet the large demands of growth, these carbon compounds can instead be partitioned into the synthesis of carbon based secondary metabolites (phenolics, tannins, etc.). In environments where the resource demands for growth are met, carbon is allocated to rapidly dividing meristems (high sink strength) at the expense of secondary metabolism. Thus rapidly growing plants are predicted to contain lower levels of secondary metabolites and vice versa. In addition, the tradeoff predicted by the GDBH may change over time, as evidenced by a recent study on Salix spp. Much support for this hypothesis is present in the literature, and some scientists consider the GDBH the most mature of the plant defense hypotheses.

## Importance to Humans

### Agriculture

The variation of plant susceptibility to pests was probably known even in the early stages of agriculture in humans. In historic times, the observation of such variations in susceptibility have provided solutions for major socio-economic problems. The grape phylloxera was introduced from North America to France in 1860 and in 25 years it destroyed nearly a third (100,000 km$^2$) of the French grape yards. Charles Valentine Riley noted that the American species *Vitis labrusca* was resistant to *Phylloxera*. Riley, with J. E. Planchon, helped save the French wine industry by suggesting the grafting of the susceptible but high quality grapes onto *Vitis labrusca* root stocks. The formal study of plant resistance to herbivory was first covered extensively in 1951 by Reginald (R.H.) Painter, who is widely regarded as the founder of this area of research, in his book *Plant Resistance to Insects*. While this work pioneered further research in the US, the work of Chesnokov was the basis of further research in the USSR.

Fresh growth of grass is sometimes high in prussic acid content and can cause poisoning of grazing livestock. The production of cyanogenic chemicals in grasses is primarily a defense against herbivores.

The human innovation of cooking may have been particularly helpful in overcoming many of the defensive chemicals of plants. Many enzyme inhibitors in cereal grains and pulses, such as trypsin inhibitors prevalent in pulse crops, are denatured by cooking, making them digestible.

It has been known since the late 17th century that plants contain noxious chemicals which are avoided by insects. These chemicals have been used by man as early insecticides; in 1690 nicotine was extracted from tobacco and used as a contact insecticide. In 1773, insect infested plants were treated with nicotine fumigation by heating tobacco and blowing the smoke over the plants. The flowers of *Chrysanthemum* species contain pyrethrin which is a potent insecticide. In later years, the applications of plant resistance became an important area of research in agriculture and plant breeding, particularly because they can serve as a safe and low-cost alternative to the use of pesticides. The important role of secondary plant substances in plant defense was described in the late 1950s by Vincent Dethier and G.S. Fraenkel. The use of botanical pesticides is widespread and notable examples include Azadirachtin from the neem (*Azadirachta indica*), d-Limonene from Citrus species, Rotenone from *Derris*, Capsaicin from chili pepper and Pyrethrum.

Natural materials found in the environment also induce plant resistance as well. Chitosan derived from chitin induce a plant's natural defense response against pathogens, diseases and insects including cyst nematodes, both are approved as biopesticides by the EPA to reduce the dependence on toxic pesticides.

The selective breeding of crop plants often involves selection against the plant's intrinsic resistance strategies. This makes crop plant varieties particularly susceptible to pests unlike their wild relatives. In breeding for host-plant resistance, it is often the wild relatives that provide the source of resistance genes. These genes are incorporated using conventional approaches to plant breeding, but have also been augmented by recombinant techniques, which allow introduction of genes from completely unrelated organisms. The most famous transgenic approach is the introduction of genes from the bacterial species, *Bacillus thuringiensis*, into plants. The bacterium produces proteins that, when ingested, kill lepidopteran caterpillars. The gene encoding for these highly toxic proteins, when introduced into the host plant genome, confers resistance against caterpillars, when the same toxic proteins are produced within the plant. This approach is controversial, however, due to the possibility of ecological and toxicological side effects.

## Pharmaceutical

Illustration from the 15th-century manuscript *Tacuinum Sanitatis* detailing the beneficial and harmful properties of Mandrakes

Many currently available pharmaceuticals are derived from the secondary metabolites plants use to protect themselves from herbivores, including opium, aspirin, cocaine, and atropine. These chemicals have evolved to affect the biochemistry of insects in very specific ways. However, many of these biochemical pathways are conserved in vertebrates, including humans, and the chemicals act on human biochemistry in ways similar to that of insects. It has therefore been suggested that the study of plant-insect interactions may help in bioprospecting.

There is evidence that humans began using plant alkaloids in medical preparations as early as 3000 B.C. Although the active components of most medicinal plants have been isolated only recently (beginning in the early 19th century) these substances have been used as drugs throughout the human history in potions, medicines, teas and as poisons. For example, to combat herbivory by the larvae of some Lepidoptera species, Cinchona trees produce a variety of alkaloids, the most familiar of which is quinine. Quinine is extremely bitter, making the bark of the tree quite unpalatable, it is also an anti-fever agent, known as Jesuit's bark, and is especially useful in treating malaria.

Throughout history mandrakes (*Mandragora officinarum*) have been highly sought after for their reputed aphrodisiac properties. However, the roots of the mandrake plant also contain large quantities of the alkaloid scopolamine, which, at high doses, acts as a central nervous system depressant, and makes the plant highly toxic to herbivores. Scopolamine was later found to be medicinally used for pain management prior to and during labor; in smaller doses it is used to prevent motion sickness. One of the most well-known medicinally valuable terpenes is an anticancer drug, taxol, isolated from the bark of the Pacific yew, *Taxus brevifolia*, in the early 1960s.

### Biological Pest Control

Repellent companion planting, defensive live fencing hedges, and "obstructive-repellent" interplanting, with host-plant resistance species as beneficial 'biological control agents' is a technique in biological pest control programs for: organic gardening, wildlife gardening, sustainable gardening, and sustainable landscaping; in organic farming and sustainable agriculture; and in restoration ecology methods for habitat restoration projects.

## Inducible Plant Defenses Against Herbivory

Plants and herbivores have co-evolved together for 350 million years. Plants have evolved many defense mechanisms against insect herbivory. Such defenses can be broadly classified into two categories: (1) permanent, constitutive defenses, and (2) temporary, inducible defenses. Both types are achieved through similar means but differ in that constitutive defenses are present before an herbivore attacks, while induced defenses are activated only when attacks occur. In addition to constitutive defenses, initiation of specific defense responses to herbivory is an important strategy for plant persistence and survival.

### Benefits of Induced Defences

Inducible defenses allow plants to be phenotypically plastic. This may confer an advantage over constitutive defenses for multiple reasons. First, it may reduce the chance that attacking insects adapt to plant defenses. Simply, inducible defenses cause variations in the defense constituents of a plant, thereby making the plant a more unpredictable environment for insect herbivores. This variability has an important effect on the fitness and behaviour of herbivores. For example, the study of herbivory on radish (*Raphanus sativus*) by the cabbage looper caterpillar (*Trichoplusia ni*) demonstrated that the variation of defensive chemicals (glucosinolates) in *R. sativus*, due to induction, resulted in a significant decrease in the pupation rates of *T. ni*. In essence, defensive chemicals can be viewed as having a particular dosage-dependent effect on herbivores: it has little detrimental effect on herbivores when present at a low or moderate dose, but has dramatic effects at higher concentrations. Hence, a plant which produces variable levels of defensive chemicals is better defended than one that always produces the mean level of toxin.

Second, synthesizing a continually high level of defensive chemicals renders a cost to the plant. This is particularly the case where the presence of herbivorous insects is not always predictable. For example, the production of nicotine in cultivated tobacco (*Nicotiana tabacum*) has a function in plant defence. *N. tabacum* plants with a higher constitutive level of nicotine are less susceptible to insect herbivory. However, *N. tabacum* plants that produce a continually high level of nicotine flower significantly later than plants with lower levels of nicotine. This results suggest that there is a biosynthetic cost to constantly producing a high level of defensive chemicals. Inducible defences are advantageous as they reduce the metabolic load on the plant in conditions where such biological chemicals are not yet necessary. This is particularly the case for defensive chemicals containing nitrogen (e.g. alkaloids) as if the plant is not being attacked it is able to divert more nitrogen to producing rubisco and will therefore be able to grow faster and produce more seeds.

In addition to chemical defenses, herbivory can induced physical defenses, such as longer thorns, or indirect defenses, such as rewards for symbiotic ants.

## Cost of Induced Defences

Central to the concept of induced defences is the cost involved when stimulating such defences in the absence of insect herbivores. After all, in the absence of cost, selection is expected to favour the most defended genotype. Accordingly, individual plants will only do so when there is a need to. The cost of induced defences to a plant can be quantified as the resource-based trade-off between resistance and fitness (allocation cost) or as the reduced fitness resulting from the interactions with other species or the environment (ecological cost).

Allocation cost is related to the channelling of a large quantity fitness-limited resources to from resistance traits in plants. Such resources might not be quickly recycled and thus, are unavailable for fitness-relevant process such as growth and reproduction. For instance, herbivory on the broadleaf dock (*Rumex obtusifolius*) by the green dock beetle (*Gastrophysa viridula*) induces an increased activity in cell wall-bound peroxidase. The allocation of resources to this increased activity results in reduced leaf growth and expansion in *R. obtusifolius*. In the absence of herbivory, inducing such a defence would be ultimately costly to the plant in terms of development.

Ecological cost results from the disruption of the many symbiotic relationships that a plant has with the environment. For example, jasmonic acid can be used to simulate an herbivore attack on plants and thus, induce plant defences. The use of jasmonic acid on tomato (*Lycopersicon esculentum*) resulted in plants with fewer but larger fruits, longer ripening time, delayed fruit-set, fewer seeds per plant and fewer seeds per unit of fruit weight. All these features play a critical role in attracting seed dispersers. Due to the consequences of induced defences on fruit characteristics, *L. esculentum* are less able to attract seed dispersers and this ultimately results in a reduced fitness.

### Sensing Herbivory Attack

Induced defences require plant sensing the nature of injury, such as wounding from herbivore attack as opposed to wounding from mechanical damage. Plants therefore use a variety of cues, including the sense of touch, and salivary enzymes of the attacking herbivore. For example, in a study to test whether plants can distinguish mechanical damage from insect herbivory attack, Korth and Dixon (1997) discovered that the accumulation of induce defence transcription products occurred more rapidly in potato (*Solanum tuberosum L.*) leaves chewed on by caterpillars than in leaves damaged mechanically. Distinct signal transduction pathway are activated in response either to insect damage or mechanical damage in plants. While chemicals released in wounding responses are the same in both cases, the pathway in which they accumulate are separate. Not all herbivore attack begins with feeding, but with insects laying eggs on the plant. The adults of butterflies and moths (order Lepidoptera), for example, do not feed on plants directly, but lay eggs on plants which are suitable food for their larva. In such cases, plants have been demonstrated to induce defences upon contact from the ovipositing of insects.

### A mechanism of Defence Induction: Changes in Gene Transcription Rates

Systemically induced defences are at least in some cases the result of changes in the transcription rates of genes in a plant. Genes involved in this process may differ between species, but common to all plants is that systemically induced defences occur as a result of changes in gene expression. The changes in transcription can involve genes which either do not encode products involved in insect resistance, or are involved in general response to stress. In cultivated tobacco (*Nicotiana tobacum*) photosynthetic genes are down-regulated, while genes directly involved in defences are up-regulated in response to insect attack. This allows more resources to be allocated to producing proteins directly involved in the resistance response. A similar response was reported in *Arabidopsis* plants where there is an up-regulation of all genes that are involved in defence. Such changes in the transcription rates are essential in inducing a change in the level of defence upon herbivory attack.

### Classification of Induced Genes

Not all up-regulated genes in induced defences are directly involved in the production of toxins. The genes encoding newly synthesised proteins after a herbivory attack can be categorised based on the function of their transcriptional products. There are three broad classification categories: defence genes, signalling pathway genes and rerouting genes. The transcription of defensive gene produces either proteins that are directly involved in plant defence such as proteinase inhibitors or are enzymes that are essential for the production of such proteins. Signalling pathway genes are involved in transmitting the stimulus from the wounded regions to organs where defence genes are transcribed. These genes are essential in plants due to the constraints in the vascular systems of the plants. Finally, rerouting gene are responsible in allocating resources for metabolism from primary metabolites involved in photosynthesis and survival to defence genes.

## Plant Use of Endophytic Fungi in Defense

Plant use of endophytic fungi in defense occurs when endophytic fungi, which live symbiotically with the majority of plants by entering their cells, are utilized as an indirect defense against herbivores. In exchange for carbohydrate energy resources, the fungus provides benefits to the plant which can include increased water or nutrient uptake and protection from phytophagous insects, birds or mammals. Once associated, the fungi alter nutrient content of the plant and enhance or begin production of secondary metabolites. The change in chemical composition acts to deter herbivory by insects, grazing by ungulates and/or oviposition by adult insects. Endophyte-mediated defense can also be effective against pathogens and non-herbivory damage.

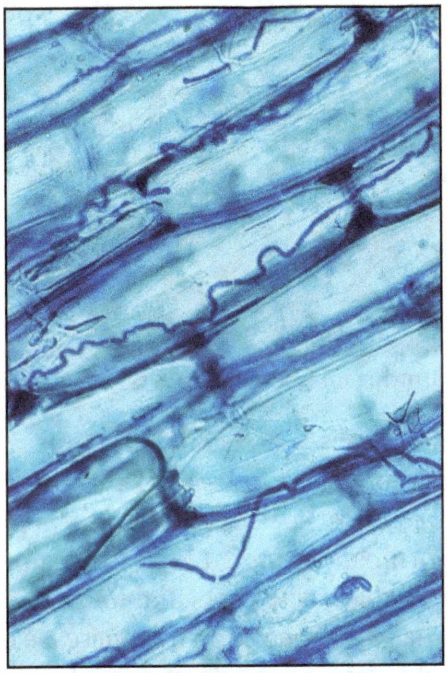

Neotyphodium spp.are commonly associated with tall fescue in the leaf sheath tissue. They produce secondary metabolites toxic to herbivores.

This differs from other forms of indirect defense in that the fungi live within the plant cells and directly alter their physiology. In contrast, other biotic defenses such as predators or parasites of the herbivores consuming a plant are normally attracted by volatile organic compounds (known as semiochemicals) released following damage or by food rewards and shelter produced by the plant. These defenders vary in the time spent with the plant: from long enough to oviposit to remaining there for numerous generations, as in the ant-acacia mutualism. Endophytic fungi tend to live with the plant over its entire life.

## Diversity of Endophytic Associations

The fungal endophytes are a diverse group of organisms forming associations almost ubiquitously throughout the plant kingdom. The endophytes which provide indirect defense against herbivores may have come from a number of origins, including mutualistic root endophyte associations and the evolution of entomopathogenic fungi into plant-associated endophytes. The endomycorrhiza, which live in plant roots, are made up of five groups: arbuscular, arbutoid, ericoid, monotropoid, and orchid mycorrhizae. The majority of species are from the phylum Glomeromycota with the ericoid species coming from the Ascomycota, while the arbutoid, monotropoid and orchid mycorrhizae are classified as Basidiomycota. The entomopathogenic view has gained support from observations of increased fungal growth in response to induced plant defenses and colonization of plant tissues.

Claviceps spp. fungus growing on wheat spikes, a common endophyte of the grasses.

Examples of host specialists are numerous – especially in temperate environments – with multiple specialist fungi frequently infecting one plant individual simultaneously. These specialists demonstrate high levels of specificity for their host species and may form physiologically adapted host-races on closely related congeners. *Piriformospora indica* is an interesting endophytic fungus of the order Sebacinales, the fungus is capable of colonising roots and forming symbiotic relationship with every possible plant on earth . *P. indica* has also been shown to increase both crop yield and plant defence of a variety of crops(barley, tomato, maize etc.) against root-pathogens. However, there are also many examples of generalist fungi which may occur on different hosts at different frequencies (e.g. Acremonium endophytes from five subgenera of Festuca) and as part of a variety of fungal assemblages. They may even spread to novel, introduced plant species. Endophytic mutualists associate with species representative of every growth form and life history strategy in the grasses and many other groups of plants. The effects of associating with multiple strains or species of fungus at once can vary, but in general, one type of fungus will be providing the majority of benefit to the plant.

## Mechanisms of Defense

### Secondary Metabolite Production

Some chemical defenses once thought to be produced by the plant have since been shown to be synthesized by endophytic fungi. The chemical basis of insect resistance in endophyte-plant defense mutualisms has been most extensively studied in the perennial ryegrass and three major classes of secondary metabolites are found: indole diterpenes, ergot alkaloids and peramine. Related compounds are found across the range of endophytic fungal associations with plants. The terpenes and alkaloids are inducible defenses which act similarly to defensive compounds produced by plants and are highly toxic to a wide variety of phytophagous insects as well as mammalian herbivores. Peramine occurs widely in endophyte-associated grasses and may also act as a signal to invertebrate herbivores of the presence of more dangerous defensive chemicals. Terpenoids and ketones have been linked to protection from specialist and generalist herbivores (both insect and vertebrate) across the higher plants.

Generalist herbivores are more likely than specialists to be negatively affected by the defense chemicals that endophytes produce because they have, on average, less resistance to these specific, qualitative defenses. Among the chewing insects, infection by mycorrhizae can actually benefit specialist feeders even if it negatively affects generalists. The overall pattern of effects on insect herbivores seems to support this, with generalist mesophyll feeders experiencing negative effects of host infection, although phloem feeders appear to be affected little by fungal defenses.

Secondary metabolites may also affect the behaviour of natural enemies of herbivorous species in a multi-trophic defense/predation association. For instance, terpenoid production attracts natural enemies of herbivores to damaged plants. These enemies can reduce numbers of invertebrate herbivores substantially and may not be attracted in the absence of endophytic symbionts. Multi-trophic interactions can have cascading consequences for the entire plant community, with the potential to vary widely depending on the combination of fungal species infecting a given plant and the abiotic conditions.

## Altered Nutrient Content

Due to the inherently nutrient-exchange based economy of the plant-endophyte association, it is not surprising that infection by fungi directly alters the chemical composition of plants, with corresponding impacts on their herbivores. Endophytes frequently increase apoplastic carbohydrate concentration, altering the C:N ratio of leaves and making them a less efficient source of protein. This effect can be compounded when the fungus also uses plant nitrogen to form N-based secondary metabolites such as alkaloids. For example, the thistle gall fly (*Urophora cardui*) experiences reduced performance on plants infected with endophytic fungi due to the decrease in N-content and ability to produce large quantities of high-quality gall tissue. Additionally, increased availability of limiting nutrients to plants improves overall performance and health, potentially increasing the ability of infected plants to defend themselves.

## Impacts on Host Plants

## Herbivory Prevention

Studies of fungal infection consistently reveal that plants with endophytes are less likely to suffer substantial damage, and herbivores feeding on infected plants are less productive. There are multiple modes through which endophytic fungi reduce insect herbivore damage, including avoidance (deterrence), reduced feeding, reduced development rate, reduced growth and/or population growth, reduced survival and reduced oviposition. Vertebrate herbivores such as birds, rabbits and deer show the same patterns of avoidance and reduced performance. Even below-ground herbivores such as nematodes and root-feeding insects are reduced by endophyte infection. The strongest evidence for anti-herbivore benefits of fungal endophytes come from studies of herbivore populations being extirpated when allowed to feed only on infected plants. Examples of local extinction have been documented in crickets, larval armyworms and flour beetles.

Yet chemical defenses produced by fungal endophytes are not universally effective, and numerous insect herbivores are unaffected by a given compound at one or more life history stages; larval stages are often more susceptible to toxins than adults. Even endophytes which purportedly provide some defense benefit to their hosts such as the Neotyphidium partner of many grass species in the alpine tundra do not always lead to avoidance or ill-effects on herbivores due to spatial variation in levels of consumption.

## Mutualism-pathogenicity Continuum

Not all endophytic symbioses confer protection from herbivores – only some species associations act as defense mutualisms. The difference between a mutualistic endophyte and a pathogenic one can be indistinct and dependent on interactions with other species or environmental conditions. Some fungi which are pathogens in the absence of herbivores may become beneficial under high levels of insect damage, such as species which kill plant cells in order to make nutrients available for their own growth, thereby altering nutritional content of leaves and making them a less desirable foodstuff. Some endomycorrhizae may provide defense benefits but at the cost of lost reproductive potential by rendering grasses partially sterile with their own fungal reproductive structures taking precedence. This is not unusual among fungi, as non-endophytic plant pathogens have similar conditionally beneficial effects on defense. Some species of endophyte may be beneficial for the plants in other ways (e.g. nutrient and water uptake) but will provide less benefit as a plant receives more damage and not produce defensive chemicals in response. The effect of one fungus on the plant can be altered when multiple strains of fungi are infecting a given individual in combination.

Some endomycorrhizae may actually promote herbivore damage by making plants more susceptible to it. For example, some oak fungal endophytes are positively correlated with the levels of damage from leaf miners (*Cameraria spp.*), although negatively correlated with number of larvae present due to a reduction of oviposition on infected plants, which partially mitigates the higher damage rate. This continuum between mutualism and pathogenicity of endophytic fungi has major implications for plant fitness depending on the species of partners available in a given environment; mutualist status is conditional in a way similar to pollination and can shift from one to the other just as frequently.

## Fitness and Competitive Ability

Fungal endophytes which provide defensive services to their host plants may exert selective pressures favouring association through enhanced fitness relative to uninfected hosts. The fungus Neotyphodium spp. infects grasses and increases fitness under conditions with high levels of interspecific competition. It does this through a combination of benefits including anti-herbivore defenses and growth promoting factors. The customary assumption that plant growth promotion is the main way fungal mutualists improve fitness under attack from herbivores is changing; alteration of plant chemical composition and induced resistance are now recognized as factors of great importance in improving competitive ability and fecundity. Plants undefended by chemical or physical means at certain points in their life histories have higher survival rates when infected with beneficial endophytic fungi. The general trend of plants infected with mutualistic fungi outperforming uninfected plants under moderate to high herbivory exerts selection for higher levels of fungal association as herbivory levels increase. Unsurprisingly, low to moderate levels of herbivore damage also increases the levels of infection by beneficial endophytic fungi.

In some cases the symbiosis between fungus and plant reaches a point of inseparability; fungal material is transmitted vertically from the maternal parent plant to seeds, forming a near-obligate mutualism. Because seeds are an important aspect of both fecundity and competitive ability for plants, high germination rates and seedling survival increase lifetime fitness. When fitness of plant and fungus become tightly intertwined, it is in the best interest of the endophyte to act in a manner beneficial to the plant, pushing it further toward the mutualism end of the continuum. Such effects of seed defense can also occur in dense stands of conspecifics through horizontal transmission of beneficial fungi. Mechanisms of microbial association defense, protecting the seeds rather than the already established plants, can have such drastic impacts on seed survival that they have been recognized to be an important aspect of the larger 'seed defence theory'.

## Climate Change

The range of associated plants and fungi may be altered as climate changes, and not necessarily in a synchronous fashion. Plants may lose or gain endophytes, with as yet unknown impacts on defense and fitness, although generalist species may provide indirect defense in new habitats more often than not. Above-ground and below-ground associations can be mutual drivers of diversity, so altering the interactions between plants and their fungi may also have drastic effects on the community at large, including herbivores. Changes in distribution may bring plants into competition with previously established local species, making the fungal community – and particularly the pathogenic role of fungus – important in determining outcomes of competition with non-native invasive species. As carbon dioxide levels rise, the amplified photosynthesis will increase the pool of carbohydrates available to endophytic partners, potentially altering the strength of associations. Infected C3 plants show greater relative growth rate under high $CO_2$ conditions compared to uninfected plants, and it is possible that the fungi drive this pattern of increased carbohydrate production.

Levels of herbivory may also increase as temperature and carbon dioxide concentrations rise. However, should plants remain associated with their current symbiotic fungi, evidence suggests that the degree of defense afforded them should not be altered. Although the amount of damage caused by herbivores frequently increases under elevated levels of atmospheric $CO_2$, the proportion of damage remains constant when host plants are infected by their fungal endophytes. The change in Carbon-Nitrogen ratio will also have important consequences for herbivores. As carbohydrate levels increase within plants, relative nitrogen content will fall, having the dual effects of reducing nutritional benefit per unit biomass and also lowering concentrations of nitrogen-based defenses such as alkaloids.

## History of Research

## Early Recognition

The effects of endophytic fungi on the chemical composition of plants have been known by humans for centuries in the form of poisoning and disease as well as medicinal uses. Especially noted were impacts on agricultural products and livestock. Recognition and study of the mutualism did not begin in earnest until the 1980s when early studies on the impacts of alkaloids on animal herbivory confirmed their importance as agents of deterrence. Biologists began to characterize the diversity of endophytic mutualists through primitive techniques such as isozyme analysis and measuring the effects of infection on herbivores. Basic descriptive accounts of these previously neglected species of fungus became a major goal for mycologists, and a lot of research focus shifted to associates of the grass family (Poaceae) in particular, because of the large number of species which represent economically important commodities to humans.

## Recent Advances and Future Directions

In addition to continuing descriptive studies of the effects of infection by defense mutualist endophytes, there has been a sharp increase in the number of studies which delve further into the ecology of plant-fungus associations and especially their multi-trophic impacts. The processes by which endophytic fungi alter plant physiology and volatile chemical levels are virtually unknown, and limited current results show a lack of consistency under differing environmental conditions, especially differing levels of herbivory. Studies comparing the relative impacts of mutualistic endophytes on inducible defenses and tolerance show a central function of infection in determining both responses to herbivore damage.[100] On the whole, molecular mechanisms behind endophyte-mediated plant defense has been an increasing focus of research over the past ten years.

Since the beginning of the biotechnology revolution, much research has been also focused on using genetically modified endophytes to improve plant yields and defensive properties. The genetic basis of response to herbivory is being explored in tall fescue, where it appears the production of jasmonic acid may play a role in downregulation of the host plant's chemical defense pathways when a fungal endophyte is present. In some cases, fungi that are closely associated with their hosts have transferred genes for secondary metabolite production to the host genome, which could help to explain multiple origins of chemical defenses within the phylogeny of various groups of plants. This represents an important line of inquiry to pursue, especially in regards to understanding the chemical pathways that can be utilized in biotechnological applications.

## Importance to Humans

## Agriculture and Livestock

The secondary chemicals produced by endophytic fungi when associated with their host plants can be very harmful to mammals including livestock and humans, causing more than 600 million dollars in losses due to dead livestock every year. For example, the ergot alkaloids produced by Claviceps spp. have been dangerous contaminants of rye crops for centuries. When not lethal, defense chemicals produced by fungal endophytes may lead to lower productivity in cows and other livestock feeding on infected forage. Reduced nutritional quality of infected plant tissue also lowers the performance of farm animals, compounding the effect of reduced feed uptake when provided with infected plant matter. Reduced frequency of pregnancy and birth has also been reported in cattle and horses fed with infected forage. Consequently, the dairy and meat-production industries must endure substantial economic losses.

Fungal resistance to herbivores represents an environmentally sustainable alternative to pesticides that has experienced reasonable success in agricultural applications. The organic farming industry has embraced mycorrhizal symbionts as one tool for improving yields and protecting plants from damage. Infected crops of soybean, ribwort plantain, cabbage, banana, coffee bean plant and tomato all show markedly lower rates of herbivore damage compared to uninfected plants. Endophytic fungi show great promise as a means of indirect biocontrol in large-scale agricultural applications. The potential for biotechnology to improve crop populations through inoculation with modified fungal strains could reduce toxicity to livestock and improve yields of human-consumed foods. The endophyte, either with detrimental genes removed or beneficial new genes added, is used as a surrogate host to transform the crops genetically. An endophyte of ryegrass has been genetically transformed in this way and used successfully to deter herbivores.

Understanding how to mediate top-down effects on crop populations caused by the enemies of herbivores as well as bottom-up effects of chemical composition in infected plants has important consequences for the management of agricultural industries. The selection of endophytes for agricultural use must be careful and consideration must be paid to the specific impacts of infection on all species of pest and predators or parasites, which may vary on a geographic scale. The union of ecological and molecular techniques to increase yield without sacrificing the health of the local or global environment is a growing area of research.

## Pharmaceutical

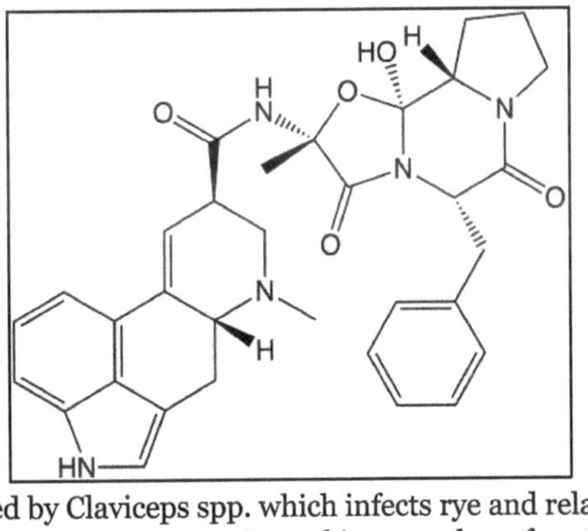


Ergotamine, a mycotoxin produced by Claviceps spp. which infects rye and related grasses, causing poisoning of livestock and humans. It is used in a number of medicinal

Many secondary metabolites from endophyte-plant interactions have also been isolated and used in raw or derived forms to produce a variety of drugs treating many conditions. The toxic properties of ergot alkaloids also make them useful in the treatment of headaches and throughout the process of giving birth by inducing contractions and stemming hemorrhages. Drugs used to treat Parkinson's Disease have been created from isolates of ergot toxins, although health risks may accompany their use. Ergotamine has also been used to synthesize lysergic acid diethylamide because of its chemical similarity to lysergic acid. The generally chemically-based defense properties of endophytic fungi make them a perfect group of organisms to search for new antibiotic compounds within, as other fungi have in the past yielded such useful drugs as penicillin and streptomycin and plants use their antibiotic qualities as a defense against pathogens.

## References

- Norman M. Wereley; Janet M. Sater (2012). Plants and Mechanical Motion: A Synthetic Approach to Nastic Materials and Structures. DEStech Publications, Inc. ISBN 978-1-60595-043-3
- Dov Koller; Elizabeth Van Volkenburgh (15 January 2011). THE RESTLESS PLANT. Harvard University Press. pp. 18–. ISBN 978-0-674-05943-6

- Black, M.H. & Halmer, P. (2006). The encyclopedia of seeds: science, technology and uses. Wallingford, UK: CABI. pp. 226. ISBN 978-0-85199-723-0
- Peterson, R.L.; Massicotte, H.B. & Melville, L.H. (2004). Mycorrhizas: anatomy and cell biology. National Research Council Research Press. ISBN 978-0-660-19087-7
- Michael Smith, C. (2005). Plant Resistance to Arthropods: Molecular and Conventional Approaches. Berlin: Springer. ISBN 1-4020-3701-5

# Permissions

All chapters in this book are published with permission under the Creative Commons Attribution Share Alike License or equivalent. Every chapter published in this book has been scrutinized by our experts. Their significance has been extensively debated. The topics covered herein carry significant information for a comprehensive understanding. They may even be implemented as practical applications or may be referred to as a beginning point for further studies.

We would like to thank the editorial team for lending their expertise to make the book truly unique. They have played a crucial role in the development of this book. Without their invaluable contributions this book wouldn't have been possible. They have made vital efforts to compile up to date information on the varied aspects of this subject to make this book a valuable addition to the collection of many professionals and students.

This book was conceptualized with the vision of imparting up-to-date and integrated information in this field. To ensure the same, a matchless editorial board was set up. Every individual on the board went through rigorous rounds of assessment to prove their worth. After which they invested a large part of their time researching and compiling the most relevant data for our readers.

The editorial board has been involved in producing this book since its inception. They have spent rigorous hours researching and exploring the diverse topics which have resulted in the successful publishing of this book. They have passed on their knowledge of decades through this book. To expedite this challenging task, the publisher supported the team at every step. A small team of assistant editors was also appointed to further simplify the editing procedure and attain best results for the readers.

Apart from the editorial board, the designing team has also invested a significant amount of their time in understanding the subject and creating the most relevant covers. They scrutinized every image to scout for the most suitable representation of the subject and create an appropriate cover for the book.

The publishing team has been an ardent support to the editorial, designing and production team. Their endless efforts to recruit the best for this project, has resulted in the accomplishment of this book. They are a veteran in the field of academics and their pool of knowledge is as vast as their experience in printing. Their expertise and guidance has proved useful at every step. Their uncompromising quality standards have made this book an exceptional effort. Their encouragement from time to time has been an inspiration for everyone.

The publisher and the editorial board hope that this book will prove to be a valuable piece of knowledge for students, practitioners and scholars across the globe.

# Index

## R



## S



## T



## V

www.ingramcontent.com/pod-product-compliance
Lightning Source LLC
LaVergne TN
LVHW080711210926
848485LV00003B/450